W0262051

PROTOPLASMATOLOGIA
HANDBUCH
DER PROTOPLASMAFORSCHUNG

HERAUSGEGEBEN VON

L. V. HEILBRUNN UND F. WEBER
PHILADELPHIA   GRAZ

MITHERAUSGEBER

W. H. ARISZ-GRONINGEN · H. BAUER-WILHELMSHAVEN · J. BRACHET-BRUXELLES · H. G. CALLAN-ST. ANDREWS · R. COLLANDER-HELSINKI · K. DAN-TOKYO · E. FAURÉ-FREMIET-PARIS · A. FREY-WYSSLING-ZÜRICH · L. GEITLER-WIEN · K. HÖFLER-WIEN · M. H. JACOBS-PHILADELPHIA · D. MAZIA-BERKELEY · A. MONROY-PALERMO · J. RUNNSTRÖM-STOCKHOLM · W. J. SCHMIDT-GIESSEN · S. STRUGGER-MÜNSTER

BAND II

CYTOPLASMA

B

CHEMIE

2

SPEZIELLE CYTOCHEMIE

b

ORGANISCHE VERBINDUNGEN

δ

CHEMISTRY AND BIOLOGY OF THE STARCH GRANULE

SPRINGER-VERLAG WIEN GMBH

1959

CHEMISTRY AND BIOLOGY OF THE STARCH GRANULE

BY

N. P. BADENHUIZEN

DEPARTMENT OF BOTANY, UNIVERSITY OF THE WITWATERSRAND, JOHANNESBURG

WITH 44 FIGURES

SPRINGER-VERLAG WIEN GMBH

1959

ISBN 978-3-211-80522-0   ISBN 978-3-7091-3849-6 (eBook)
DOI 10.1007/978-3-7091-3849-6

Chemistry and Biology of the Starch Granule

By

N. P. BADENHUIZEN

Department of Botany, University of the Witwatersrand, Joh. nesburg

With 44 Figures

Contents

I. Introduction

The study of the starch granule concerns biologists and chemists alike, and therefore it is continuously necessary to synthesize the results of both. Neglect of one or the other has led to extreme confusion in the past and the development of fantastic theories. With the advance of our knowledge it has become quite clear that the chemical analysis of isolated starch is one thing, but that synthesis and breakdown of starch in the living cell is quite another. In fact, we still know very little about the latter, which is directly linked with fundamental problems of cell differentiation. In most cases we cannot do very much with the chemical results when we try to apply them to the biology of the living cell. We meet here with intriguing problems, some of which are outlined below.

The starch granules as stored in various plants, have been described often (for drawings see e. g. Gassner 1955). This article emphasizes the unusual and some less well known facts, because the deviation from the normal often opens the way to the solution of a problem.

It is essential that our ideas about the structure of the starch granules are critically reviewed from time to time. New tools and more careful experiments add continuously to our knowledge. Radioactive carbon and the electron microscope recently gave direct information about the development of starch granules. As in the case of cellulose, these methods were able to confirm conclusions already obtained from indirect evidence, but at the same time the interpretations often require the utmost care.

The factors determining shape, structure and chemical composition of the starch granules belong to the most fundamental problems that require our attention. Genetics and the physiological environment of the cell are fields of study which will become more important in the future (BADEN-HUIZEN 1954).

We know that certain genes influence starch content and the percentage of linear molecules. We know very little about the interaction of the various enzymes involved and the structure of the plastids influencing starch granule structure.

The field to be covered is extremely wide, and therefore recent literature will be discussed mainly. Papers which give useful summaries in various fields have been marked with an asterisk in the list of references.

A number of new observations are included in the present paper.

II. The Natural Starch Granule
A. General Considerations
1. Microscopical

One difficulty encountered in microscopic studies is the fact that the starch granule swells to a certain extent when immersed in water, and then becomes transparent. Dry, concentric starch granules of wheat or rye swell about 30% in diameter, dry eccentric granules of potato or pea show differential swelling power along the two axes. Measuring the longitudinal axis potato granules showed an increase of 47%, pea granules a shrinkage of 2%; the short axis showed an increase in both cases of 29% and 35% respectively (HANSSEN et al. 1953). At the same time cracks and a layered structure may become visible and the consistency changes from hard to soft, as we find when we cut the granules with a micro-knife (BADENHUIZEN 1938, 1946). When investigated in pentane, which has the same refractive index as water but prevents swelling, no such structures as mentioned above are apparent (HANSSEN et al. 1953).

We may therefore well ask whether these structures are not artefacts. The cracks certainly are, characteristic as they may be e. g. for rye and bean starch (Fig. 18 and 19), but what about the so-called layers?

During further swelling at higher temperatures other structures may appear, referred to as "blocklets." Especially when starch granules show reduced swelling power, either naturally (cereal starches) or as an acquired character following certain treatments, it looks as if the layers are broken up regularly into small parts, the blocklets (Fig. 21). Are they preformed structures of the native granule?

During the final stages of swelling most starch granules develop a sac filled with a colloidal solution (Fig. 13). Is the wall of this sac derived from a special outer membrane of the original granule?

These examples make it clear that it will often be difficult to distinguish between essential structures and artefacts. We do not see structures in the dry granules, but as soon as we cause swelling to occur we change the original conditions, and structures may appear which only reflect the real structure. Secondly, the granules become transparent, and as a consequence, what we see is an "optical section", a projection of three-dimensional structures on to the plane of focus (Fig. 3).

It is indeed very difficult to build up a three-dimensional picture from such optical sections. No wonder that many conflicting theories have been put forward in the past, and are still being made in the present. Fig. 20 is an attempt to present a model in space, combining the effect of projection, but will be discussed below. It needs the hand of an artist to draw structures in swollen starch granules in their right perspective.

2. Chemical

The situation often becomes even more confused when structures, whether artefacts or not, are related to chemical composition. Before 1940 most results tended to show a chemical homogeneity in starch. Microscopical observations could only show the equality in structure of the different layers of a granule, but of course could not have a decisive influence (Badenhuizen 1938).

Later the presence of *linear molecules (amylose)* and *branched molecules (amylopectin)* was demonstrated very convincingly (Meyer 1952), although French investigators (Badenhuizen 1949, p. 68–69) did not accept this multiple concept. Even now an isolated voice (Bauer 1953) upholds the "unitary theory," maintaining that the starch granule is one big molecule, and that amylose and amylopectin are artefacts resulting from chemical hydrolysis.

It is, in fact, not so easy to decide which concept is the correct one, from making chemical analyses. Recently Ulmann (1956) tried to decide the issue by using a chromatographic method of separation. When a starch solution is forced through an Al_2O_3-column, of which the upper part is acid (pH 4.5) and the lower part is alkaline (pH 8.2), amylopectin is retained in the acid region, and amylose appears as a band from pH 7 onwards. Ulmann's recommendation is to grind the granules, after which they become soluble in cold water. Such a solution gave the same chromatographic results as a boiled one, and therefore would support the multiple concept. We know however that mechanical damage eventually results in depolymerisation (Lampitt et al. 1948), so that it cannot be a criterium for or against the multiple concept.

A strong argument for the pre-existence of the two fractions in starch may come from the fact that granules grow by apposition (well-established for potato starch; Badenhuizen and Dutton 1956), which means that layers are deposited on the outside, often at irregular time intervals, probably depending on a varying carbohydrate supply. It is difficult to imagine how the molecular chains of the new layer could link up with those of the old layer, especially as the latter contains starch in the retrograded condition, when it can not fulfil the function of primer molecules needed for synthesis by phosphorylase (Whelan and Bailey 1954). Further the unitary theory would make it even more difficult to explain why we always find a genetically fixed and practically constant percentage of linear molecules in a particular starch (Table 1), than it is now.

Table 1. *Amylose content of various starches.*

Plant	Amylose %	Plant	Amylose %
Waxy types		*Pith of trunk*	
Sorghum	0.29 (a), 1–2 (b)	Sago	23.9 (a), 26 (b, c)
Amaranthus leucosperma	1 (b)		
Maize	1 (c), 3–9 (b)	*Underground storage organs*	
Rice	0–2 (b)	Potato	22.9 (f), 21 (c), 20–24 (b)
Barley	7 (b)		
Granadilla	1 (d), 6.5 (e)	Sweet potato	20.4 (f), 17.8 (c), 20 (b)
Grasses (normal)		Tapioca	16,7 (c), 18.5 (f), 18.9 (a), 18–19 (b)
Maize	23.8 (f), 23,9 (a), 24 (c)		
		Canna	30.1 (a), 26–33 (b)
Maize vars.	20–36 (b)	*Maranta*	20.5 (c), 21 (b)
Sorghum vars.	21–28 (b)	*Iris*	27 (c), 27–29 (b)
Wheat	24.3–25.1 (a), 25 (c, g)	*Dioscorea alata*	24.8 (i)
		D. anguina	26.3 (i)
Wheat vars.	18–27 (b)	*Alstroemeria*	21 (b)
Barley	19 (o), 22 (c), 24–27 (b)	*Arisaema*	23 (b)
		Colocasia esculenta	16–17 (b)
Oat	26 (c), 23–24 (b)	*C. antiquorum*	11 (b)
Rice	18.5 (c), 16–17 (b)	*Xanthosoma*	11 (b)
Rye	26 (b)	Tulip	26–27 (b)
Coix lachryma	24–27 (b)	Easter Lily	31 (b)
Panicum miliaceum	27 (b)	Parsley	19 (b)
Sudan Grass	29–33 (b)	Parsnip	11.1 (m)
Leguminosae		*Miscellaneous seeds and fruits*	
Smooth peas	35 (h), 35.4 (a), 34–37 (b)	*Fagopyrum esculentum*	24.9 (a), 26–28 (b)
		Hevea	20 (l)
Wrinkled peas	66 (h), 64.7 (a), 61–70 (b)	Banana	16.8 (c), 19 (b)
		Avocado	31 (b)
Various beans	30–35 (b)	Cashew	23 (b)
Cicer arietinum	33 (b)	*Chenopodium* species	22–25 (b)
Lathyrus odoratus	25 (b)	Horse chestnut	24–26 (b)
Lentil	30 (b)	*Quercus* spp.	27–31 (b)
Vetch	36 (b)	*Nelumbo lutea*	31 (b)
Trees (sapwood)		*Leaves*	
Elm	20.5 (k)	Sunflower	13.8 (p)
Maple	19 (n)		

(a) LARSON and GILLIS (1953); (b) DEATHERAGE *et al.* (1955); (c) ANDERSON *et al.* (1955); (d) CILLIE and JOUBERT (1950); (e) BADENHUIZEN (1955 a); (f) DOREMUS *et al.* (1951); (g) BAUM *et al.* (1955); (h) POTTER *et al.* (1953); (i) RAO and BERI (1955); (k) CAMPBELL *et al.* 1951); (l) GREENWOOD and ROBERTSON (1954); (m) ANDERSON and GREENWOOD (1956); (n) BALLOU and PERCIVAL (1952); (o) MacWILLIAM and PERCIVAL (1951); (p) RADWAN and STOCKING (1957).

Lastly: both fractions are also found when leaching is done under mild conditions and in the absence of oxygen (Cowie and Greenwood 1957). For these and other reasons it is most likely that the two types of molecules are present in the starch granule; amylose and amylopectin are both on the market nowadays (Anonymous 1957).

The discovery that starch contains oxygen-sensitive bonds (Bottle *et al.* 1953; Baum and Gilbert 1954) made previous determinations of the molecular weight of the fractions obsolete. When fractionation takes place in the absence of oxygen, a 10 × higher average degree of polymerization (D. P.) is found for potato amylose (4000) (Cowie and Greenwood 1957), than is usually reported (Umrau and Smith 1957). The molecular weight found for any linear fraction depends upon methods used for isolation and fractionation of the starch. See Greenwood (1956) for more detailed discussions of the chemical characterization of the two fractions, and their separation.

When therefore the starch molecules are larger than was previously thought, it does not mean that they all have the same size. On the contrary, there are strong indications that both the linear molecules (Foster and Zucker 1952; Lewis and Smith 1957; Cowie and Greenwood 1957) and the branched molecules (Witnauer *et al.* 1952; Meyer and Settele 1953; Fischer and Settele 1953; Anderson and Greenwood 1955; Lewis and Smith 1957; Stacy and Foster 1957) show extreme heterogeneity.

Linear molecules of different chain length could for example be separated by the use of collodion membranes of different porosity (Mould and Synge 1952), by crystallization from aqueous pyridine (Foster and Paschall 1953), or by careful leaching (Cowie and Greenwood 1957). However, Ulmann (1954) demonstrated that fractionation may make the components more heterogeneous than they were before in the mixed (starch) solution, so that the fractions can not be used for drawing conclusions about components in native starch.

For the branched molecules the laminated or comb-like models have been generally abandoned in favour of Meyer's irregular tree-like structure (Bourne 1952; Peat *et al.* 1952; Larner *et al.* 1952; Peat *et al.* 1956). The average chain length calculated for the branched molecules from 15 different starches (data taken from Anderson *et al.* 1955) appears to be 21 glucose units, but in some starches like "amylomaize" (a new starch with high amylose content) and that of wrinkled peas a much higher figure (36) has been reported (Wolff *et al.* 1955), simulating a higher content of linear molecules than is actually present. Something similar had also been suggested for Easter Lily starch with 30–34% amylose (Stuart and Brimhall 1943). The outer branches are always longer than the inner ones; some parallel arrangement is thought to be possible between the outer branches (Fig. 1) (see also p. 36).

Spot tests have been worked out to differentiate between the two types of molecules (Nussenbaum 1951; Potter *et al.* 1953), and they can easily be recognised in swollen granules under the microscope, after staining with iodine (Czaja 1954, 1955). Ionophoretic properties of the borate complexes and their iodine-staining properties seem to be good criteria for the purity of the fractions (FOSTER *et al.* 1956). Fractions intermediate in degree of branching may occur (DOREMUS *et al.* 1951).

LANSKY *et al.* (1949) found 5–7% of corn starch to consist of slightly branched amylose, while potato amylose should have linear molecules only (also POTTER and HASSID 1951). Other investigators (e. g. KERR and CLEVELAND 1952) think that potato and tapioca amyloses have a few branching points, and that corn amylose is entirely linear.

This is an important point, but one which is evidently far from being settled. Fractionation may take place at a certain degree of branching (BADENHUIZEN 1951), and it would then be impossible to distinguish between linear and weakly branched molecules. The lack of intermediate zones on a chromatographic column (ULMANN 1954) could indicate that there are no intermediate fractions at all. They may be simulated by the impurity of the fractions (BECHTEL 1951), or by the occurrence of long outer branches. It has not yet been possible to arrive at a satisfactorily fractionation for barley starch (MAC WILLIAMS and PERCIVAL 1951; BENGOUGH *et al.* 1957).

In the linear parts of the molecules glucose units are linked by a-1,4 glucosidic linkages; at the branching points a-1,6 linkages have been demonstrated. The evidence for the possible occurrence of a-1,3 linkages has been summarised by GIRI (1957); we cannot, however, exclude the possibility that they are artefacts caused by reversion (CORI 1956). The presence of a-1,3 linkages is based on the production of nigerose after hydrolysis (PEAT *et al.* 1957; WOLFROM and THOMPSON 1956, 1957) under conditions in which reversion is said to be negligible. Models of molecules show that such linkages would be sterically possible (ANONYMOUS 1954). WHELAN and ROBERTS (1952) found fructose in glycogen and waxy maize starch!

All these questions are not of a purely chemical interest. They are vital to our understanding of the structure and the growth of starch granules, and therefore it is of the greatest importance that we should have a more definite knowledge about them.

3. Physico-chemical

Starch granules are birefringent. As they consist for about 75% of branched molecules, the parallel arrangement of outer branches (Fig. 1) should be responsible for the birefringence in the first place. It is then not surprising that "waxy" varieties, which contain branched molecules only, have the same intensity of birefringence as normal types (BAKER and WHELAN 1950). The birefringence of cereal starch however is generally lower than that of tuber starch, which in turn is much lower than that of cellulose.

Birefringence is caused by the radial arrangement of optically positive chains, the regularity of which is demonstrated by the polarisation cross. This disappears after mechanical deformation of the granule, e. g. in a ball mill, although the persistence of the birefringence still indicates a strong association between molecular branches.

Linear molecules seem to play a subordinate role. They mainly act as "reinforcing rods"; the more there are the stronger the net-work is (KITE *et al.* 1957; see p. 19). Still it would seem strange that the birefringence of wrinkled pea starch, with its high amylose content, long outer branches, and B-spectrum (see p. 12), is less intensive than that of maize starch. Baker and Whelan (1950) suggest that this may be due to incomplete orientation as a result of rapid crystallisation; such starches should be more easily mobilised. This may be true for leaf starch (Radwan and Stocking 1957), but it could hardly apply to wrinkled peas, in which the granules are very resistant.

It has been pointed out earlier that molecular association is stronger in cereal than in "tuber" starch, for which several reasons can be given (Badenhuizen 1951). In addition corn amylose appears to have an unusually high tendency to aggregation (Foster and Paschall 1953).

Fig. 1. "Crystalline" amylopectin.

In general we see that stronger molecular association, as brought about by various factors, depresses birefringence, and this is contrary to what one would expect.

In some starch granules of a high amylose maize variety (genotype *aesu*) birefringence practically disappears (Fig. 40). The gene *su* appears to be mainly responsible for this effect (Dvonch *et al.* 1951), which is mainly due to irregular deposition (p. 55).

If we exclude a possible contributing influence of proteins in wrinkled pea starch, or phospholipids in cereal starch, it would mean that radial orientation of linear chains becomes increasingly difficult with chain length. Long chains may be folded up in various directions when they are precipitated, and their segments can be partly associated with each other, partly with the outer branches of the amylopectin, at the same time causing the latter to deviate more or less from the radial direction. In other words, the "reinforcing rods" are crooked.

Stacy and Foster (1956) have pointed out the limited deformability of branched molecules, against the high deformability of long linear chains. The application of "shear" brings the chain segments of amylopectins closer together, which results in orientation favourable for secondary bonding, followed by aggregation and precipitation; linear molecules on the other hand are stretched by the same process.

Now the stronger the molecular association, whatever the causes, the greater will be its disturbing influence on general radial orientation, and birefringence will decrease accordingly.

Where there is a high content of linear chains, radial arrangement may become distorted to the extent that birefringence disappears altogether. In all cases linear molecules appear to contribute to the structure of local crystalline regions, which therefore are mixed crystallites (Meyer and Menzi 1953). The separate deposition of linear and branched molecules. as advocated by Nikuni (1957) is extremely unlikely.

In cellulose the linear glucose chains are stretched and ordered in parallel fashion, except in bacterial cellulose and in that of *Ascidia*, which seem to be branched structures (VAN DER WYK and SCHMORAK 1953). It would be interesting to compare birefringence in these cases. Branching causes the birefringence of potato starch to be about 25% of that of plant cellulose.

About 25 years ago, KATZ and co-workers discovered that natural starch could crystallize in two main patterns, as followed from a study of the X-ray diffraction spectra (KATZ and v. ITALLIE 1930, SAMEC and KATZ 1933). They distinguished an A-spectrum (characteristic for starch from grasses) from a B-spectrum, whilst a C-spectrum could be explained as being a mixture of A- and B-patterns (see Table 2 on p. 12). These different crystalline patterns are interconvertible. Evaporation of a solution of soluble potato (B-) starch at room temperature gives a product with B-spectrum, but above 60° C. an A-spectrum appears. Autoclaving potato starch with little moisture gives it an A-spectrum and other characteristics of cereal starch (WHITTENBERGER and NUTTING 1948).

The unit cell of B-starch is larger than that of A-starch, and therefore in the latter the molecules are lying more closely to-

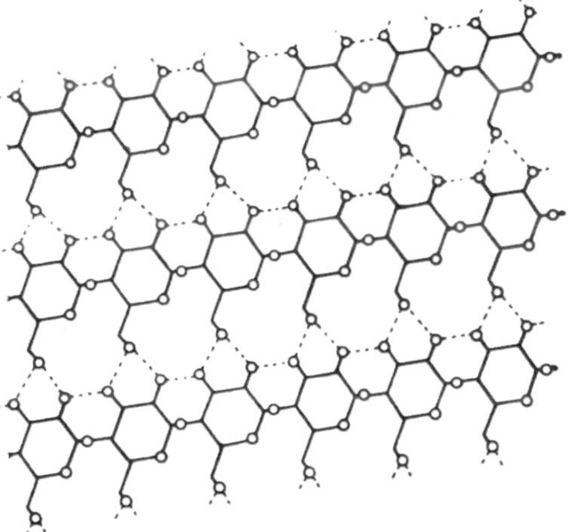

Fig. 2. Suggested pattern of bonds and hydrogen bonds in the V-modification of starch. The planar arrangement pictured is to be thought of as bent around a cylinder in such a way as to produce a continuous helical chain molecule, with adjacent turns connected by hydrogen bonds.
(After HUGGINS 1957; courtesy J. Chem. Educ.)

gether. B-starch is more hydrated than A-starch (HELLMAN *et al.* 1954), potato starch being the most hygroscopic of all (OHASHI *et al.* 1957).

Since we know that high amylose maize starch, and also wrinkled pea starch, shows a B-spectrum, it has become clear that the crystalline pattern has no direct relation to the strength of molecular association. B-starch can have weak (potato) or strong molecular association (wrinkled pea). The type of crystalline pattern is probably determined by the amount of water available during precipitation, little water giving an A-spectrum. Only when the amount of linear molecules is the same in both cases, can we say that A-starch has stronger molecular association and reduced swelling power, as compared to B-starch, and this could explain the difference in birefringence between maize and potato starch.

From ULMANN's (1954) work it would follow that the water of crystallization, held by one glucose unit, may vary from 1 to 5 molecules, with water content in the granules varying from about 10 to 35%. This water is necessary for the orientation of chains in crystalline patterns. Anhydrous or strongly hydrated starches are amorphous. Molecular association is brought about by H-bonds, in the first place between the primary OH-groups of the 6th C-atom in the glucose units (Fig. 2) because of their steric freedom (Anonymous 1954). This is the case when sufficient water is available. As less water becomes available during crystallization, more H-bonds will be formed between secondary OH-groups, especially at position 2 and 3 of adjacent rings, and the chains will approach each other more closely. At the same time we observe transitions from the B- to the A-spectrum.

A strong association between molecules may easily cause tensions, which can be released in the form of miniature cracks whenever the slightest swelling takes place. Ultrathin sections of such granules would show a rugged surface in the electron microscope, corresponding to the "microgranules" observed by Nikuni (1957). Sections through potato starch granules have a smooth texture, as we would expect. The "micro-granular" structure is probably related to the phenomena discussed above, and it will be interesting to see whether it exists e. g. in starch of wrinkled peas, and whether it is generally absent in B-starches with low amylose content. This rough inner structure is probably of more importance in regard to enzymic digestion than just the surface structures (Thornburg 1956), which moreover are reported to be relatively smooth (Whistler *et al.* 1955).

The degree of molecular association is at the basis of most of the differences between cereal and other starches. Transitory stages, showing a C-spectrum, do occur (banana—Subrahmanyan *et al.* 1957; *Curcuma, Zingiber*—Badenhuizen 1939) and additional factors may have some influence, but in general molecular association determines swelling power, digestibility, and other properties of starches.

The presence of complex-building organic substances like *fatty acids*. alcohols, etc. does not appear to have an influence on the distance between starch molecules (Mikus *et al.* 1946), but they definitely promote retrogradation.

The influence of *proteins* is not clear. Thirty years ago Samec assumed the existence of an amylopectin-protein complex in wheat starch (see Badenhuizen 1937, p. 307), while recently Rozenfeld and Physhevskaya (1954) reported that proteins can form complexes, especially with the branched molecules. It looks as if the proteins are adsorbed on to the surface of the carbohydrate particles. For whole granules this was found to be the case for starch from *Hevea* seeds (Greenwood and Robertson 1954), while it was actually observed for wheat starch in electron photographs (Hess *et al.* 1955).

Both the nature of the protein and particle size (specific surface) may be involved. Schwimmer and Balls (1949) found that α-amylase was adsorbed by preference and better by A- and C-starches than by B-starches. Fractionation improved the adsorption capacity, whereby in this case amylose was far more active than amylopectin. On the other hand freshly precipitated Lintner starch and glycogen were the most efficient of all.

The general conclusion was that the smaller particles had a better adsorption capacity for α-amylase.

Gelatinised starch may become so "built in" by protein molecules that they are no longer accessible to amylase activity (WUNDERLY 1940). Intact starch granules are relatively non-porous (HELLMAN and MELVIN 1950) and proteins will not penetrate (LATHE and RUTHVEN 1956), but as soon as their structure is damaged, adsorption is improved (SCHWIMMER and BALLS 1949).

It follows that the colloid-chemical structure of the starch and the protein composition of the stroma of the amyloplast may be important factors influencing the development of the starch granule. We do not know enough about the former, and nothing about the latter factor.

B. The Layers

One of the conspicuous features of many starch granules is their layering. Many theories have been put forward to explain the occurrence of these layers (BADENHUIZEN 1937), but for potato starch in any case we can be sure that they do not bear any correlation to the alternation between day and night (ROBERTS and PROCTOR 1954; MES and MENGE 1954; BÜNNING and HESS 1954). Of course the layers are optical sections of transparent shells which envelop each other (Fig. 3). It has been demonstrated for potato tubers, that ^{14}C-labelled starch can be deposited around the starch granules, which therefore grow by apposition (BADENHUIZEN and DUTTON 1956).

Layering is much less conspicuous for instance in cereal starch, but in most cases it can easily be made visible by means of lintnerization (prolonged treatment with $7\frac{1}{2}\%$ HCl). This is possible because there is an alternation of layers which are rich in crystalline starch with layers which are amorphous and rich in water (HESS 1955). The first are birefringent, the latter are seen as black lines between crossed nicols. Any chemical alteration happens in the first place in the less resistent amorphous layers, and that is also the reason why lintnerization accentuates layering (Fig. 4), showing that it is a topochemical reaction and not mainly a surface reaction, as is sometimes maintained (BAUER 1953; KNYAGINICHEV et al. 1955). However, penetration is time-dependent and therefore the *initial* attack takes place at the periphery (COWIE and GREENWOOD 1957). Acid-hydrolysed, thin-boiling starches show an increase in apparent linear material (CALDWELL 1952), and in general it is found that acid breaks down the branched molecules first (DIMMICK and CORLEY 1950; ULMANN 1956; COWIE and GREENWOOD 1957). Birefringence remains unchanged, but with the relative increase of linear material, swelling power decreases, and the structures eventually become very brittle (BADENHUIZEN 1938, 1946). Very rightly COWIE and GREENWOOD (1957) draw the conclusion from these observations, that there cannot be a chemical difference between amorphous and crystalline layers: they both contain linear and branched molecules. There is no indication whatsoever that would justify CZAJA's (1954) assumption of alternate layers of amylose and amylopectin.

The difference is mainly one of hydration and accessibility. We therefore can expect that all crystalline layers will behave in a similar way when subjected to agents which cause their swelling, and this has been entirely confirmed by microscopical observations (Badenhuizen 1938).

Electron pictures of sections through potato starch granules show lamellar rings with a spacing of 2–7 μ, whereas in maize starch the spacing of such rings is 0.5–1 μ (Whistler and Turner 1955). In the latter the layers are therefore closer together and consequently adhere somewhat more strongly, and that is one reason why they are less conspicuous. A strong adherence between crystalline layers is also found in banana-, *Curcuma-, Maranta-* and some other starches (Badenhuizen 1938) which we indicated above as having an intermediate position as far as molecular association is concerned. We further draw attention to the fact that such starches also display an intermediate X-ray diffraction pattern, viz. the C-spectrum (see Table 2).

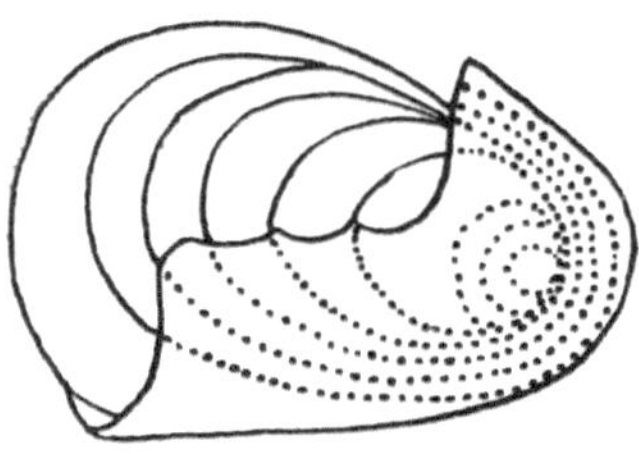

Fig. 3. Lintnerized potato starch granule with one broken shell, showing the others in optical section.

It is interesting to note from Table 2 that B- and C-starches are found in various plant organs, so that only the A-starch forms a well-defined group in this respect, as its spectrum is characteristic for grasses (except high amylose maize).

Table 2. *X-ray diffraction patterns of various starches*

A-spectrum	B-spectrum	C-spectrum
Wheat	Chestnut (seed)	Arrow-root (rhizome)
Rice	Sweet chestnut (seed)	Tapioca (tuber)
Glutinous rice	Granadilla (fruit)	Banana (fruit)
Maize	*Canna* (rhizome)	Sago (stem pith)
Waxy maize	Potato (tuber)	*Curcuma* (rhizome)
Waxy barley	*Lilium longiflorum* (bulb)	Arum lilies (tuber)
Oat	*Oxalis latifolia* (bulb)	*Colchicum* (tuber)
	Pellia (seta)	Sweet potato (tuber)
	High amylose maize	Buckwheat (seeds)
	Wrinkled pea	(Peas), beans, lentils (seeds)
	Smooth pea var. (Greenfeast)	Radish (tuber)
	All retrograded starch	
	Stale bread	
	Synthetic amylose	

The diffraction patterns are further entirely independent of size and shape of the granules, as found in different species.

Table 2 has been composed from the one published by Samec and Katz (1933), and from observations made by other investigators. How difficult it is to generalize, follows from the fact that Katz listed peas under the column C-starch, whereas the writer found starch from the Greenfeast variety to have a B-spectrum.

As an A-spectrum is found for both normal and waxy maize, the percentage of linear molecules does not have any influence on the type of crystalline pattern, which appears to be determined in the first instance by the branched molecules. This is confirmed by what is found for starches with high amylose content. In such starches the inner and outer branches of the amylopectin fraction are longer than usual, making the branched fraction one which is intermediate in its properties between amylose and normal amylopectin (WOLFF *et al.* 1955). For some types of high amylose maize starch, DVONCH *et al.* (1951) had already found that *they* gave a B-pattern. A B-spectrum was also found for starch from *ae ae* and *ae su* maize with 65% amylose (GAFIN 1957). These are first instances of

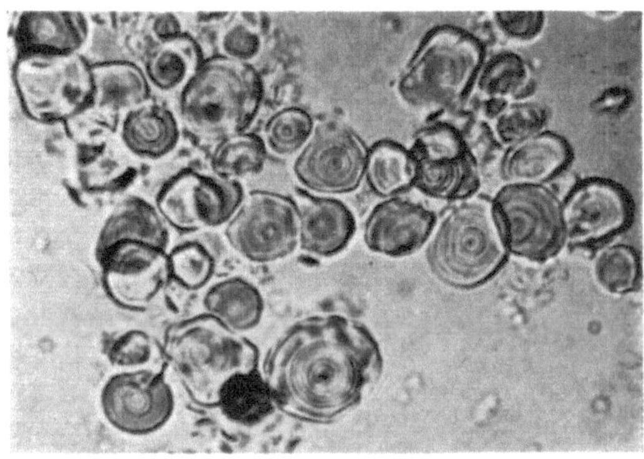

Fig. 4. Lintnerized starch granules of waxy maize.

B-spectra occurring in starch from grasses (see p. 12). Other B-starches, like potato, show us that the length of the branches is not the determining factor for the type of spectrum (Table 3 on p. 18). In this respect Granadilla starch is of particular interest. It consists mainly of amylopectin, with free branches containing only 17 glucose units (CILLIE and JOUBERT 1950); nevertheless it shows a B-spectrum (GAFIN 1957).

Molecular association is playing a leading role in the distinction between the two types of layer. In fact, waxy maize starch shows beautiful layers after lintnerization (Fig. 4), while it becomes increasingly difficult to demonstrate them in normal maize starch, wrinkled pea (BADEN-HUIZEN 1955 a, b) and high-amylose corn starch (Fig. 5).

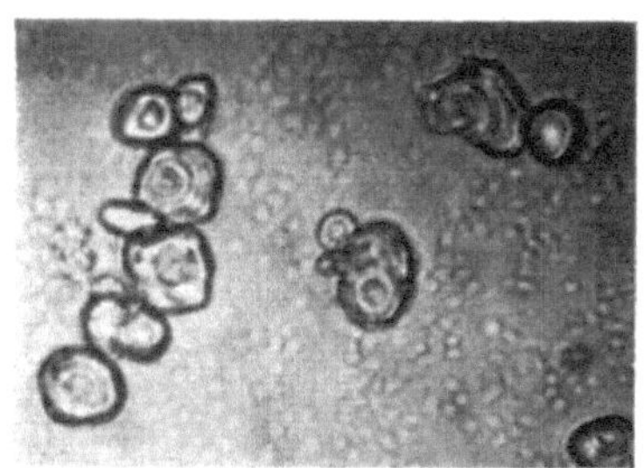

Fig. 5. Granules from high-amylose (*aeae*) maize starch after treatment with saliva at 40° C.

Ultrathin sections through maize endosperm show the presence of layers in less than 15% of the starch granules, and the explanation given is that they are the only ones which had their amyloplast intact during development (THORNBERG 1956). It strikes one, however, that layers may be visible in parts only of such sections (Fig. 6, which has been drawn after Fig. 6 in NIKUNI and WHISTLER 1957; see also the right granule in Fig. 8 of WHISTLER and THORNBERG 1957), and we may ask whether the absence of

Fig. 6. Section through a starch granule of maize; the section is folded and shows layering in one part only. (After NIKUNI and WHISTLER 1957.)

layers in so many sections of corn starch is perhaps a consequence of a disturbance of the structure caused by the cutting. A section through a granule of potato starch, reproduced by Nikuni and Hizukuri (1957) shows no sign of layering, although we know this to be conspicuous in the intact granule. In this connection it is significant that 0.1 μ thick sections do not show birefringence, while sections 1 μ in thickness do (Whistler and Thornberg 1957). Sections of 0.25 μ do not necessarily have a uniform thickness, and where the paracrystalline pattern is destroyed, birefringence disappears and the layers become invisible. That they were there nevertheless follows from sections showing corrosion canals (Fig. 7), and from the fact that after sufficient lintnerization all granules (whether normal: A. Meyer 1895, p. 194, or waxy: Badenhuizen 1955) show conspicuous layering. We therefore have to be very careful in interpreting such phenomena in terms of development in the plastid.

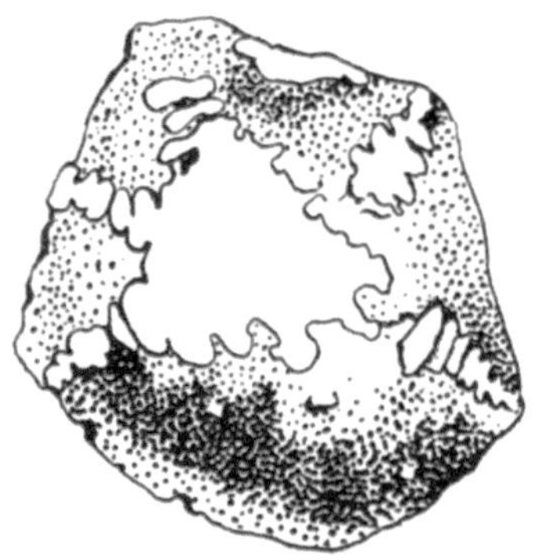

Fig. 7. Section through a corroded starch granule of maize.
(After Nikuni and Whistler 1957.)

In maize starch granules layers start becoming visible 18 days after pollination (Thornberg 1956), i. e. after they have reached a certain size.

In general it was found that layers were more often seen in large granules than in small ones (Whistler and Thornberg 1957), although the authors could not offer an explanation for this observation. The explanation however is given by the difference in swelling power between adjacent crystalline layers, and between granules of various sizes in one species (p. 18). The fundamental principle here is that volume increases more rapidly than surface. This causes layers to slide along each other when they swell, as has been demonstrated for fibers (van der Houven van Oordt-Hulshof 1957). The result is a dislocation of any connection that might exist between the layers, and this is equal to mechanical damage, causing starch molecules to become liberated from the crystalline pattern with consequent swelling: an amorphous layer is formed.

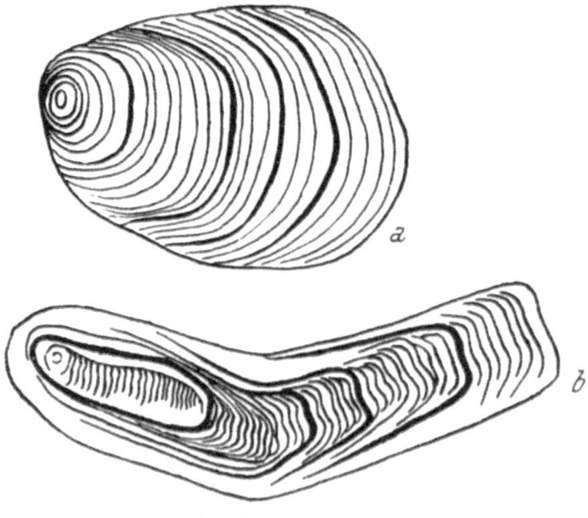

Fig. 8. Starch granules showing swelling of layers in groups.
a Oxalis ortgiesi, b Dieffenbachia seguina.
(After A. Meyer 1895.)

The connection between the layers, as they are deposited one around the other, is a loose one, but one can visualise that adhesion is brought about because molecules of the newly deposited layer partly penetrate into the existing layer. The greater the swelling, the more conspicuous the layers will be, and therefore a great amount of linear molecules, or strong molecular association in general, will suppress the effect.

Likewise layers will be better visible in larger granules (with higher swelling power).

We therefore may consider the "amorphous layers" to be artefacts, produced from material derived from the "crystalline layers," and therefore they must have the same chemical composition. In fact, it is sufficient to speak of "the layers."

The amount of starch deposited in one layer is dependent on the capacity of the amyloplast, and on the carbohydrate supply. We now *understand* fully that layers may be visible even under constant conditions. This means also that starch granules in living cells are in a swollen condition, which is confirmed by the cracks and radial striations they often show.

Further attention is drawn to the fact that the "amorphous layers" are always thin; they simply mark the "crystalline layers" by dark lines, which nevertheless can vary in width, so that some of them are broader and more conspicuous (Fig. 8). Naturally, the difference in swelling power between a layer and its adjacent one will be small, but will become much more pronounced for two layers separated by several others. As a result layers will swell in groups, but tensions between such groups are bound to increase, so that greater dislocations are produced between groups than between individual layers.

Dehydration and mechanical damage cause the "amorphous layers" to disappear, and such factors may be responsible for the disappearance of visible layering from ultra-thin sections.

C. Compound Granules

When two or more centres of crystallization are formed in a plastid, the growing granules will touch upon each other and behave as a unit. We call such granules *compound.* They may later become covered with common layers, and then they are *semi-compound* (Fig. 36). This condition is again genetically fixed, as some species are characterised by simple, and others by compound granules of constant appearance.

We know very little about the physiological conditions in the cell that cause the granules to be compound or not.

BURGEFF (1932) described highly compound (micropolyadelphous) granules in the tuberous rhizomes of a number of tropical terrestrial Orchids. The individual granules have a size varying from $0.1–2\,\mu$ and belong to the smallest known (Fig. 9). In some species one plastid may produce hundreds of them, each showing a polarization cross. Those in the centre of the aggregate are often larger than the others. The stroma of the amyloplast extends in between the individual granules and swells easily in water, causing a most peculiar disintegration of the aggregate, a process which can be retarded by using a sucrose solution. During sprouting of the tuber the amylase present in the stroma can therefore act upon the individual granules inside the aggregate, these granules becoming increasingly smaller during the process. Therefore the aggregates are highly accessible to enzymes, their starch is easily mobilised, and Burgeff points out that this condition is characteristic for rapidly-growing Orchids.

Different variations were observed by Burgeff, which are interesting to note. First of all starch granules may be compound in one organ, and simple in another, as in *Nervilia*. The large aggregates are formed by the simultaneous production of new granules, while those first deposited grow; the latter are in an eccentric position and their size is slightly larger (*Didymoplexis*). In some species one cell can contain aggregates with many small granules, or with a few larger granules (*Cystorchis*). Lastly new layers may be deposited around the aggregates, which then become semi-compound and show only one polarization cross.

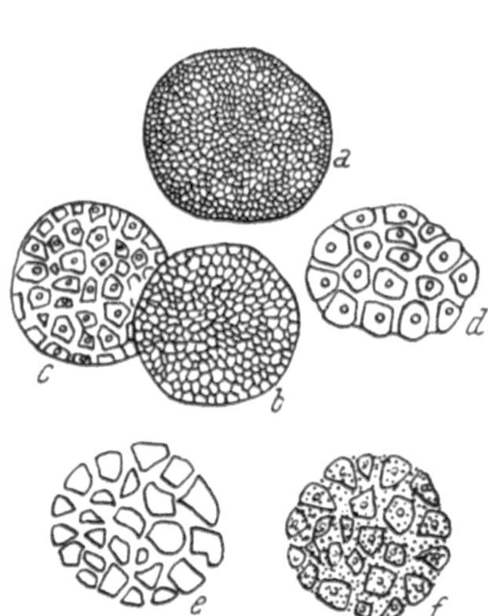

Fig. 9. Compound starch granules from *Cystorchis aphylla. a* normal, *b-f* abnormal, *e* optical section, *f* surface.
(After Burgeff 1932.)

All these variations present many intriguing problems. Burgeff himself noticed that with an excess of sugar the granules, making up the aggregate, became fewer in number and about 10 times larger than usual, a condition which was also found in *Cystorchis.*

Maige (1935) observed that compound granules were formed when pieces of *Ficaria* tubers were immersed in 5% glucose, whereas they normally formed simple starch granules. In tobacco a phenomenon has been described (Badenhuizen 1941) whereby the leaves produce yellow areas enclosed by a green network of veins ("marmer") as the result of a sudden untimely growth inhibition before maturation. In these yellow areas chloroplasts which previously had small simple granules, were transformed into amyloplasts producing large compound starch granules. Naturally, when growth stops, more sugar is available in situ for starch production. Similar effects have been obtained artificially by the application of maleic hydrazide (see also p. 55).

It looks as if sugar concentration in the cell is one of the factors controlling the production of simple or compound starch granules. A systematic investigation would be desirable.

A number of artefacts, caused by swelling, can simulate compound granules, and will be discussed in the next section.

III. Swelling Phenomena

A. Mechanism of Swelling and Influencing Factors

Any agent which is able to disrupt the H-bonds between adjacent starch molecules will cause swelling. The best swelling agents are those with a high dielectric constant, water being amongst them. Swelling is influenced by a great number of factors, e. g. granule size, percentage of linear fraction, degree of molecular association, size and homogeneity of the molecules, pretreatment, temperature, electrolytes present, extent of mechanical damage, rapidity of heating, etc. It is characteristic for a 3-dimensional molecular network.

The transition to the hydrated condition is called *gelatinization.*

A certain amount of energy is necessary to break molecular association and will therefore start where the attracting forces are weakest, i. e. in the amorphous regions of the granule. Very long linear molecules may become partly detached, but shorter molecules will diffuse out at an early stage. Cowie and Greenwood (1957) found that leaching of potato starch with water of 70⁰ C. (with the exclusion of oxygen) removed 40% of the total quantity of linear molecules with a purity of 98% and a D. P. of 1830. Dispersion of the residues gave a linear product of D. P. 5300 and a very pure branched fraction.

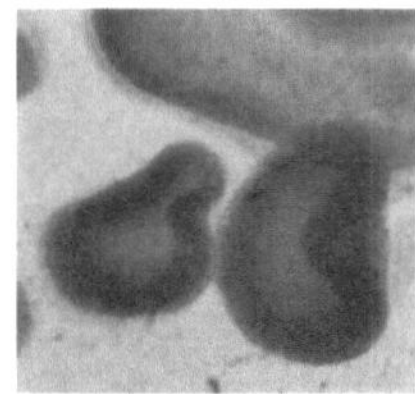

Fig. 10. Swollen sac-shaped potato starch granules after staining with iodine, showing characteristic folds.

For the preparation of a linear fraction for use in quantitative iodimetry Lambert and Rhoads (1956) extract for 2 hours at 57⁰–60⁰ C.

Swelling is dissolution delayed by cross-bonds; the more cross-bonds, the less the swelling, which becomes clear when cross-bonds are artificially introduced (Caldwell 1952, Mazurs *et al.* 1957); in this case the substance tends to become insoluble.

It is interesting that an opening of the network under the influence of acid (lintnerization, p. 11) also decreases swelling power, but the substance now becomes soluble.

With increased supply of energy gradually stronger molecular associations are disrupted, until even the most crystalline parts become swollen. The starch granules have now become little sacs (Fig. 10, 13) and together they form a paste. At high concentrations the sacs stick together because the molecules of one sac partly penetrate into the wall of another.

It follows that after a strengthening of the crystalline regions more energy is needed to overcome the intermolecular forces, in other words swelling power decreases, and a higher temperature has to be applied to reach a particular degree of swelling as compared with untreated starch. This can be done in several ways. On p. 9 the conversion of B-starch into an A-type by autoclaving with a limited amount of water was discussed. That this treatment has little effect on cereal starch (cf. Table 2) needs no explanation.

Another method is to expose starch to the prolonged action of an excess of hot water (I found best action at about 60⁰ C.). This again causes an increase in crystallinity, which is often called *retrogradation* (the word aggregation has been used especially in connection with maize starch), and again a higher temperature is required to loosen the increased number of H-bonds. For this reason starch granules gelatinize more rapidly when they are heated quickly in water, than when the temperature is raised

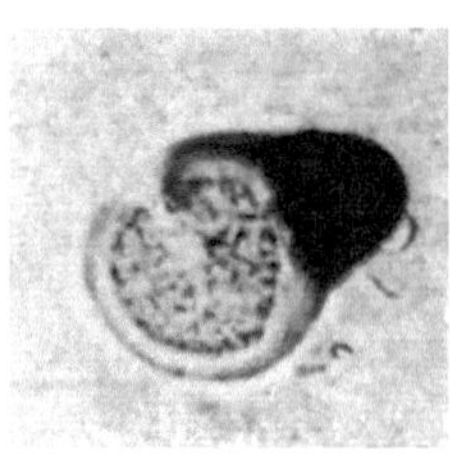

Fig. 11. Incompletely swollen starch granule of potato, the intact part stained with iodine. Inside the swollen part retrograded amylose.

slowly. As a result, gelatinization ranges have a comparative value only, and are strictly dependent on the conditions chosen. Generally the disappearance of the polarization cross is taken as a criterion for gelatinization, but it is clear that there is no really exact criterion. Starch granules of one sample vary in swelling power, each having its own gelatinization temperature; individual granules may even swell in stages (Fig. 11). Because the largest granules swell before the smaller ones (p. 15), the overall effect is that of a temperature range, which is characteristic for the type of starch. For *the* gelatinization temperature we can use that temperature at which 50% of the granules are swollen (Schoch and Maywald 1956), or that temperature at which the greatest number of granules swell at the same time (Badenhuizen 1948,

Table 3

Starch	Birefringence end-point in centigrades	X-ray spectrum	Average percentage amylose	Average chain length of amylopectin
Maize, normal	69.5	A	24	20
Maize, waxy	68.5	A	1	20
Maize, *aeae*	(90)	B	65	(36)
Maize, *aesu*	(90)	B	65	(36)
Maize, *dusu₂*	56.5	(B)	50	36
Pea, wrinkled	(90)	B	65	(36)
Potato	67	B	22	22
Tapioca	70	C	18	22

p. 217). It follows immediately that gelatinization is an irreversible process, the limit of which is determined for each granule by the temperature of the water or the concentration of the swelling agent under the conditions of the experiment (cf. Fig. 11). It has been shown for maize that different genes have a varying effect on birefringence end-point temperature (Pfahler *et al.* 1957).

In Table 3 data have been brought together from various authors, so that they are not strictly comparable. The values for birefringence end-point temperatures for example, as given by Schoch and Maywald (1956), differ considerably from those listed by Pfahler *et al.* (1957). The general trends, however, are clear. For the maize genotypes see p. 48.

The figures, placed between brackets, are guesses. It follows from the table that the amount of linear fraction bears no direct relation to the gelatinization end point, and neither does the average chain length of the amylopectin, nor the X-ray spectrum.

Some starches have a very high gelatinization temperature, and they hardly swell after boiling. They have a high amylose content, and long amylopectin branches in common. Maize genotype $dusu_2$ shows the latter two characters equally well, but has the lowest gelatinization temperature of the series. It is also clear from the examples in Table 3 that the same type of spectrum can cover a wide variety of conditions.

All we can say at present is that during crystallization unknown factors are involved which may influence molecular association. What exactly is gene su_2 doing? (See also p. 49.)

Mechanical damage disrupts the molecular network and lowers the gelatinization temperature accordingly. Although the gelatinization range is the same both for waxy and normal varieties, waxy maize starch granules swell more rapidly and more extensively, and the walls of their swollen granules are much less stable to stirring forces than those of normal maize starch, tapioca being intermediate in this respect (Mazurs *et al.* 1957).

A certain amount of water must be available for swelling to take place: there is no swelling in concentrated H_3PO_4 or NaOH. A 1.5 mol. solution of sucrose prevents potato starch from swelling at 70^0 (Täufel *et al.* 1956), because both starch and sucrose compete for water molecules, and sucrose has the greater affinity. As a result the starch lacks sufficient water for swelling. The influence of different sucrose concentrations on starch properties has been further studied by Hester *et al.* (1956).

B. Iodine and Spirals

We have seen that in living cells the starch granules are already in a swollen condition. To a certain extent there must be free chain segments with hydrated OH-groups, and such segments are most probably wound in a helicoidal shape.

Some of the arguments which support this are based upon the blue colour which starch develops in an iodine solution (for a method giving great permanency of colour, see Little 1957). Iodine is taken up in the first place by the longest linear branches, and can therefore be used to measure amylose content potentiometrically. The titration curve shows an inflection point from which the iodine binding capacity of the linear fraction can be calculated.

Retrogradation destroys the iodine complexing ability. The loss of this ability was used by Loewus and Briggs (1957) as a very sensitive method for studying retrogradation under varying conditions. A certain degree of swelling is necessary for the blue iodine colour to develop. Crystalline amylose also stains blue with iodine. A study of its dichroitic behaviour makes it probable that the molecules have a spiral shape: the X-ray diffraction pattern is a so-called V-spectrum (Rundle and French 1943).

The general conclusions are that the blue iodine colour develops only when spirally wound chains and both iodine and triiodide ions are available, and that the iodine molecules are arranged inside such spirals (Lambert 1951); and further that there is a spectrum characteristic for such helicoidal chains, to which Katz gave the designation of V-spectrum (Badenhuizen 1937) (Fig. 2).

Because starch stains blue with iodine solution, it must contain a number of spiral chains, and therefore be partly in a swollen condition. Chains which are liberated from a crystalline pattern, tend to assume a helicoidal shape and contract, while during retrogradation the reverse process takes place: the molecules are stretched.

Chains of iodine molecules are formed in general when iodine is deposited in canals, as offered by the macromolecules of starch, the crystalline pattern of α-cyclodextrins, cholic acid, etc. Boundaries between the iodine molecules disappear, and the chains of iodine molecules behave like a resonance unit with easily moving electrons. According to Cramer and Herbst (1952) this explains the deep colour of such systems, which is further dependent on the length of the chains, the degree of aggregation (Foster and Paschall 1953), and possibly the iodide concentration (Ono et al. 1953).

It must be further borne in mind that impurities like proteins (Anderson and Greenwood 1955), fatty acids (Takaoka and Nikuni 1953, Nikuni et al. 1953) and bacteria (Loewus and Briggs 1957) may interfere with the results.

It will now be clear that iodine absorption is characteristic for the type of starch investigated and that it is considerably increased when damaged granules are present (Coton et al. 1955).

Therefore the iodine binding-power of amylose depends upon the source of the amylose, the methods used for fractionation, and also on the conditions of the potentiometric titration (concentration of iodide and iodine, pH, temperature). This accounts for much variation in the amylose content of starches, as reported by various investigators.

In Table 1 on p. 5 only those results which were closely related, have been incorporated, while much of the variation is due to the different varieties studied. Whistler and Weatherwax (1948) found for primitive maize varieties a mean of 24–25%, with 22.2% and 28.3% as extremes. The extremes were 17.5 and 21.7% for sweet potatoes, with the majority between 20–21.7% (Doremus et al. 1951).

It would have been difficult, however, to include for example the values given by Singh et al. (1956), as these are consistently and markedly lower than most results of other investigators. This must be due to the conditions chosen whereby it makes a great difference whether starches are defatted or not, since fatty acids lower the affinity to iodine (Schoch 1945). The writer agrees entirely with Greenwood and Robertson (1954), when they recommend the iodine binding capacity of the pure amylose isolated from the starch under investigation, to be the only reliable basis for the estimation of the percentage of amylose in that starch. As a rule the linear fraction from potato is used as a general basis. The iodine affinities (mg. I_2 absorbed by 100 mg. of amylose) vary considerably for different amyloses: 21.2% for potato (SINGH *et al.* 1956), 19.7% for tapioca, 9.7% for Sorghum (OHASHI 1957), and for that reason no great value should be attached to the absolute values of the data in Table 1.

The iodine affinities of the branched molecules with their short free chains are now expected to be at a much lower level than those of the amyloses, and were for example found to vary between 1.2% and 0.05% (OHASHI 1957).

As a result the linear molecules stain a deep blue with iodine solution, while the amylopectins show several shades of violet. These different colours can be seen beautifully when starch is treated according to CZAJA's (1954) prescriptions. The result is a collection of brownish sacs with blue amylose inside, or diffusing out. The writer found that waxy maize starch gives only the brown-colouring sac, while Granadilla starch showed much blue amylose and thus contains rather more than 1% of the latter (BADENHUIZEN 1955 a).

As a rule the components are characterised by their so-called "Blue Value," which again has only limited value as it is dependent on the molecular weight and the method of preparation.

In lintnerized starch mainly stretched chains in crystalline patterns are left, after the acid has degraded most of the branched molecules. As a consequence iodine solution does not stain this product blue, but the colour may be yellow or pink. When the preparation is allowed to dry in the iodine (plus KI), addition of water produces a blue colour, while the granules disintegrate. The micelles have become disintegrated by the high electrolyte concentration, and in contact with water their chains become hydrated, and contract as they assume a spiral shape.

According to FREDERICK (1955) chains with less than 6 glucose units may remain straight, but above that limit they form spirals, which have the hydrophilic groups on the outside and the hydrophobic groups on the inside. This picture explains why hydrophobic substances can enter the spirals (HIRAI 1953). Apart from iodine, fatty acids, and other complex-forming substances, mention has been made of vanillin (RUTGERS 1955).

C. Swelling of Starch Granules

1. Normal swelling of B-starches

Of the B-starches potato starch is the most characteristic type, and has been studied most extensively. It has a high swelling power (12.7% when measured as linear swelling in a water-saturated atmosphere over the vacuum-dry dimension; HELLMAN *et al.* 1952), although not so great as that of tapioca or waxy maize (28.4% and 22.7%, respectively). We found a linear increase of about 15% both for small and large granules when they were transferred from pentane to water. This is a logical consequence from the fact that each layer is produced by the same process of precipitation, and has the same chemical composition and packing of molecules. The swelling power pro unit of surface is constant. In other words: a certain number of molecules in their micellar connection will give rise to a certain degree of swelling, depending on the conditions. There is no difference in the behaviour of the starch substance as such in small and large granules, and their composition is the same.

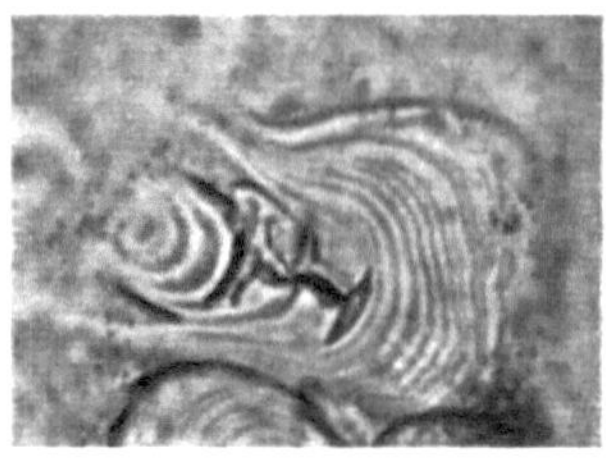

Fig. 12.   Roasted starch granule of potato.

A starch granule augments its volume as it becomes more hydrated, and a cavity develops at the hilum. This happens because we are dealing with a layered structure, in which the layers swell individually (see also p. 28).

As the radially oriented chains become more excitated, they are gradually loosened from the network. The liberated OH-groups attract water, the chains contract and become thicker. The result is that the layer swells in the tangential direction and the granule expands by *tangential swelling*. All layers swell together at the same time, and naturally a cavity will appear (Figs. 6 und 11).

Now, with increasing radius, volume increases more than surface. Therefore, even if the degree of swelling is the same for inner and outer layers, the latter will enclose a relatively larger volume than the first by the same increase in surface. For that reason they tend to separate from each other, as can be seen, when we study pyrodextrins in water (Fig. 12). During *normal* swelling layers stick together, and the result is that the outer layers extend the inner ones. The phenomena caused by this process, after slight swelling, were

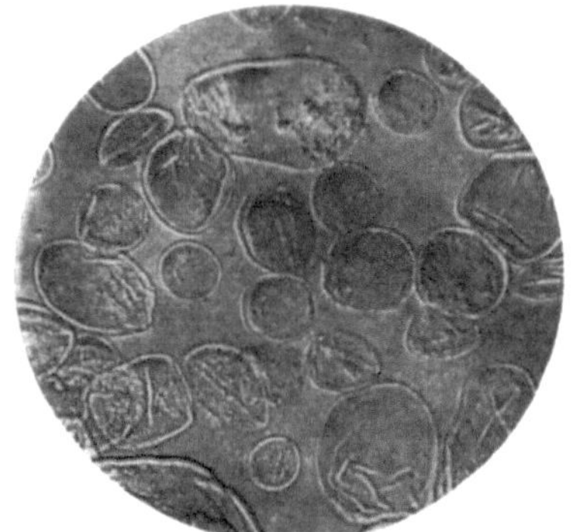

Fig. 13.   Collapsed swollen starch granules of potato.

discussed on p. 14. As swelling proceeds, the cavity is enlarged, the molecules contract and become more and more hydrated, some of them diffuse

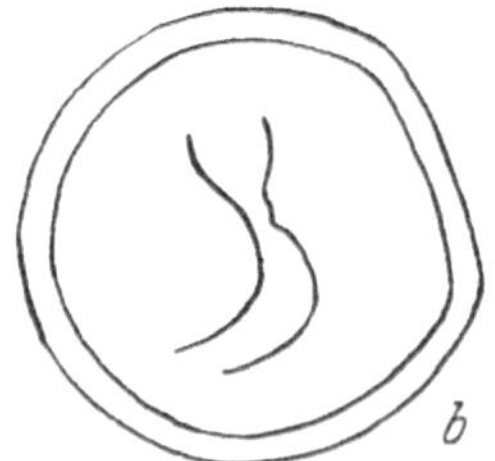

Fig. 14.   Shape of swollen starch granule of *a* waxy maize, *b* normal maize.

out, and thus the layers gradually become thinner and merge into one another. At intermediate stages (Figs. 11, 16) layers can still be counted, and even isolated (Badenhuizen 1938), but during the final stages they form the thin wall of a sac, which eventually collapses (Fig. 13).

The pull of the outer layers exerted upon the inner ones causes a more rapid disintegration of the crystalline pattern of the inner layers (Fig. 6), and this is the reason why molecules diffuse into the inside of the sac as they are liberated (Fig. 11). It has often been assumed that these molecules inside the sac would set up an osmotic pressure, which in turn would expand the sac (MEYER 1952). This would not explain the start of the swelling, nor the fact that swelling is a process limited by the conditions. Moreover, the osmotic pressure of these large molecules would be very small. Experiments have shown that the wall of the sac is not a semipermeable membrane. The inner contents are not under pressure, as can be demonstrated by micromanipulation (BADENHUIZEN 1938), and fragments of the wall show as much contraction in sugar solution, as does the whole sac (BADENHUIZEN 1953). Moreover, the shape of the sac reflects the shape

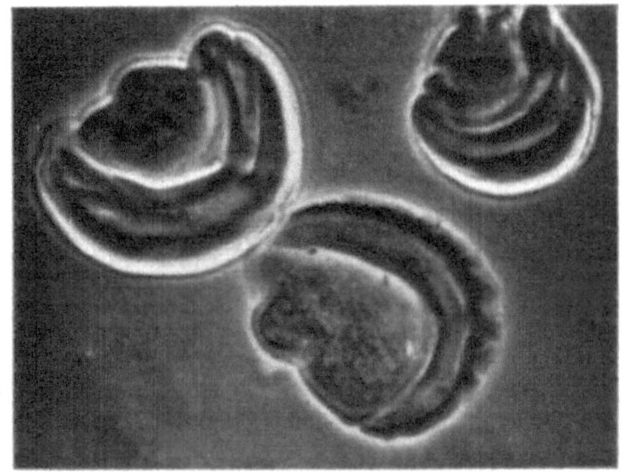

Fig. 15. Swollen starch granules of *Oxalis latifolia*. Phase contrast.

of the original starch granule (BADENHUIZEN 1938). and can be folded, as in the case of waxy maize (Fig. 14). Swelling by tangential expansion, without the interference of osmotic processes, is well established. Starch granules with low swelling power, like those of *Oxalis latifolia,* often give striking illustrations (Fig. 15). When partly gelatinized potato starch granules are kept at room temperature for several days, the amylose inside the swollen parts retrogrades, and the walls become soluble. After heating, the swollen parts dissolve, leaving open cup-shaped residues, from which the amylose contents escape. Three stages of this process are illustrated in Fig. 16 *a, b* and *c*. The confluence of the layers is especially clear in *c*. Upon

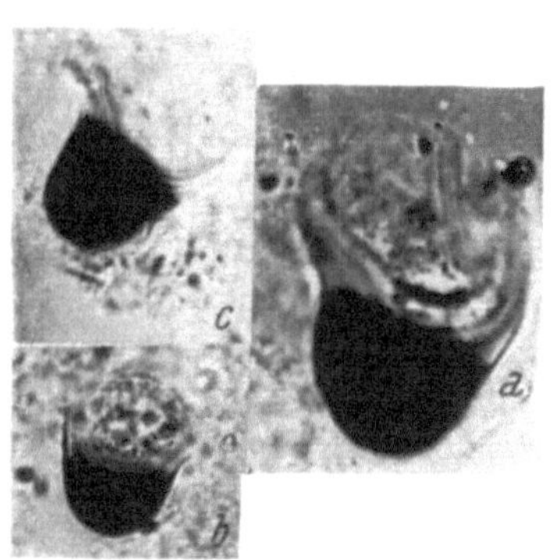

Fig. 16. Several stages in the isolation of a ball of retrograded amylose from partly gelatinized potato starch granules.

further heating the whole granule is transformed into a sac, which is now open at the top. This is a convincing demonstration of tangential swelling and lack of inner pressure (see also BADENHUIZEN 1938).

The difference in swelling pressure between large and small granules gives the former a more conspicuous layering and a slightly looser texture, so that they will swell and stain first. Tuber-starches do not develop cracks at the hilum because of their generally high swelling power.

The wall of the sac arises from the fusion of all layers (Figs. 6, 16), and is therefore not derived from a hypothetical outer membrane surrounding the original starch granule. The hypothesis of an outer membrane is one that has appealed to many investigators in the past, but all the experimental evidence is against it (ALSBERG 1938. BADENHUIZEN 1938, 1953 b).

Nevertheless, many text-books still sustain primitive ideas about the structure of the starch granule.

The correct assessment of the various stages of gelatinization is sufficient proof in itself of the non-existence of a special outer membrane. Moreover, when several of the outer layers are removed from a potato starch granule by means of saliva amylase (Fig. 17), the core left shows normal swelling upon heating (Badenhuizen 1938).

Again it is found for wheat starch, that when part of a starch granule is damaged, swollen in water, and digested with amylase, the resistance of the residue is similar to that of undamaged granules (Sandstedt 1955).

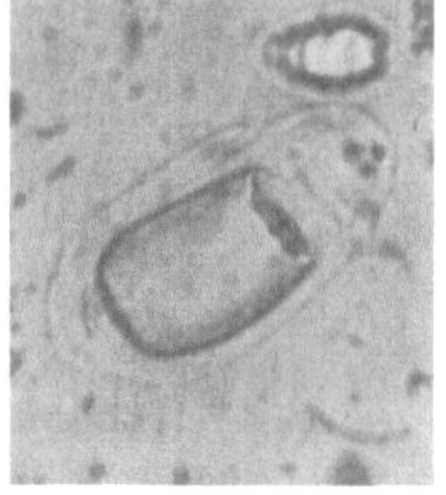

Fig. 17. Potato starch granule after treatment with saliva at 40° C. The outer layers have become transparent and will dissolve upon heating.

Recently Ulmann (1956, 1957) confirmed these observations by demonstrating chromatographically that the walls of the sacs, obtained after swelling in 30% sodium salicylate, consist of branched molecules only, whereas the solution contains the linear component. Similar results were obtained with 30% NH_4CNS, although this swelling agent dissolves some amylopectin as well. Therefore the branched fraction from all the layers is found in the wall of the sac, giving a further argument for tangential swelling and against a special outer membrane. After lintnerization Ulmann found linear and branched molecules both in the supernatant and in the residues, so that again there is no question of HCl just damaging an outer membrane (p. 11).

The molecules in the wall of the sac still have some arrangement. When granules are heated to 10° C. above gelatinization temperature and, after cooling, are brought into contact with iodine solution or ethanol, a partial return of birefringence can be observed (MacMasters 1953). This applied to A-, B- and C-starches, although the intensity of the effect was strongest in the cereal starches. The latter behaved like "root" starches after defatting, indicating that much of the strength of the molecular association in A-starches comes from the presence of complexing fatty substances.

2. Phenomena due to reduced swelling power

Reduced swelling power is found naturally in A- and C-starches, which have a higher primary resistance than most B-starches, because of a stronger molecular association (Badenhuizen 1955 b). It has an immediate effect upon the intrinsic viscosity of the boiled starches; Kawamura and Fukuba (1957) find the following values: for 8 legume starches 1.60–2.11, for potato 2.65. Swelling power can also be reduced artificially in all starches by increasing the solubility of the product (either by roasting or by lintnerization), by establishing cross-bonds between the molecules, or by retrogradation (secondary resistance). In all these cases there is no smooth swelling as described above for potato, but a number of structures develop, which are all to be considered as artefacts, although some have been interpreted in the past as being the expression of preformed structures in the granule.

In the following a few examples are given.

(a) Natural starches

As mentioned before, a number of starches have central cracks whilst still in the living cell, or they develop them when the air-dry starch is investigated in water. As such cracks develop in relation to the structure of the starch granule, they are often characteristic of a particular kind of starch. They are well-known features in bean starch (Fig. 18), and some of the larger granules of rye starch (Fig. 19). In other starches a higher degree of swelling is necessary in order to produce them.

Cereal starch granules are more rigid structures than potato starch granules, as shown by their higher gelatinization range. In the first stages of swelling they develop a system of radial fissures. In Fig. 20 a hypothetical globular granule is represented, showing in the lower half (C) how the substance produces narrow canals as it is extended, while in the left quadrant (A) pieces cut out of the layers are drawn in perspective.

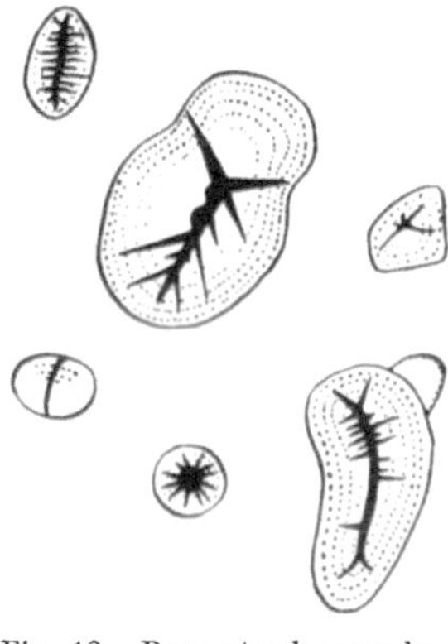

Fig. 18. Bean starch granules with characteristic fissures.

The formation of such radial canals is characteristic of A-starch (Fig. 27). The molecules have a very strong association, and they are separated in groups when tensions are produced. A beginning of this process will take place in the granules whilst they are still in the cells, and may be responsible for the "granular" structure, seen with the electron microscope in thin sections (p. 10).

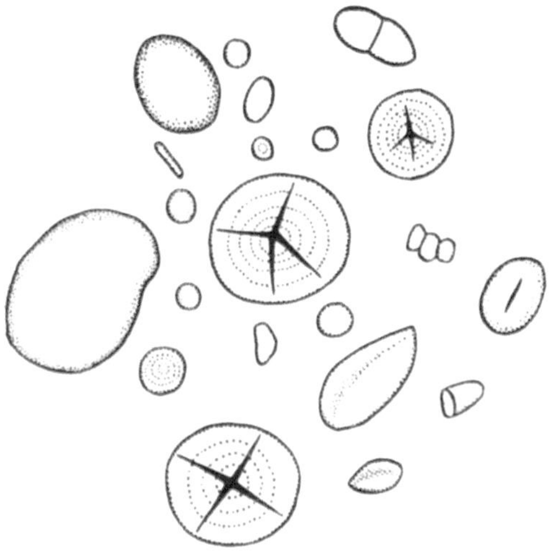

When because of the swelling, layers become more conspicuous (p. 14), the superimposing of layers and radial canals, as seen in optical section, may give the impression of the layers becoming cut up into a regular system of "blocklets." This condition, characteristic of transparent objects like starch granules, is represented in quadrant (B) of Fig. 20.

Fig. 19. Rye starch granules with characteristic fissures.

In reality the shells are torn into irregular pieces (Fig. 21). It is the optical section only which may give the impression of regularity. The pieces are certainly not of uniform size and are the result of mechanical forces.

The "blocklets" are merely artefacts, and have nothing to do with preformed structures (BADENHUIZEN 1937, 1938, 1955 b; FREY-WYSSLING 1953,

p. 317). This conclusion is supported by electron studies of sections. Whistler and Turner (1956) found for maize starch that the layers appear to be broken up tangentially into "clumps," which often show radial continuity. The smalles "clumps" of dense matter observed were 0.1–0.15 μ in width and 0.3–0.5 μ long, while the "blocklets" described by Hanson and Katz (see Badenhuizen 1937 a and b) had a size of about 1 μ. The size of the "microgranules" described by Nikuni and Hizukuri (1957) is about 200–300 Å. Blocklets and microgranules are therefore not directly comparable, but their appearance may be due to the same structural features of the starch substance. It is significant that microgranules were observed in sections of A-starches; they were not seen in potato starch, but they appeared after lintnerization, and also in this case their presence is an indication of reduced swelling power.

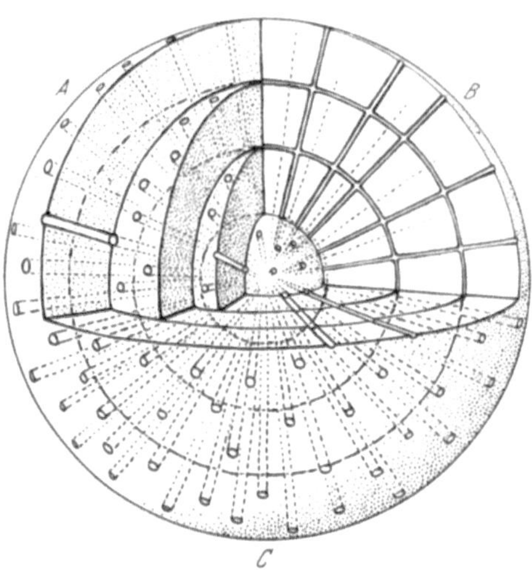

Fig. 20. Model of a hypothetical globular starch granule, showing fine radial canals after slight swelling *(C)*. In quadrant *A* pieces have been cut out so as to show the layers in perspective. Quadrant *B* is an optical section giving the impression of "blocklets."

In a beautifully illustrated paper Sandstedt (1955) describes the origin of the blocklets for rye starch granules, but he still assumes that they may be preformed structures.

Some authors thought that the "blocklets" would consist of amylose surrounded by amylopectin membranes, others assumed the reverse situation to be true. Nikuni *et al.* (1957) think that the "microgranules" visible in sections may be identical with separately precipitated amylose and amylopectin molecules. From the present description it will be clear that the "blocklets" consist of less swollen starch substance, and that they contain both components.

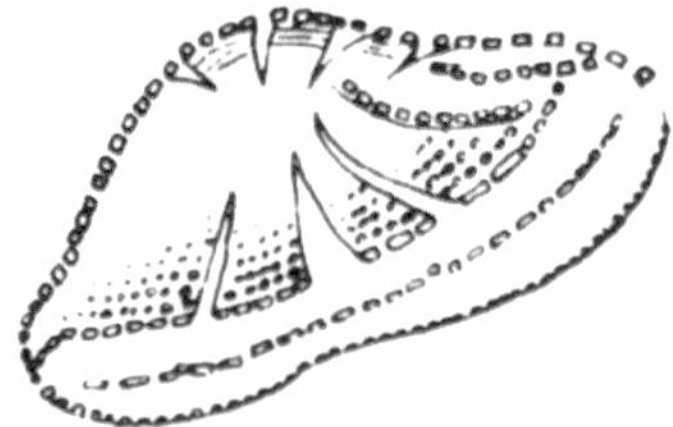

Fig. 21. Potato starch granule showing formation of "blocklets" after swelling in 0.5% NaOH.

That the chemical composition is not a factor in the appearance of "blocklets" follows from the fact that they can be clearly demonstrated in waxy maize (Badenhuizen 1955 a). Only the physical structure is of importance for them to appear (see also p. 55).

This physical structure is, in A-starches, partly determined by the presence of fatty acids (see also p. 24), which are in the first place linked to the linear fraction (they increase the viscosity of pastes only when amylose is present, Mitchell and Zillmann 1931). They can even be used for selective precipitation of this fraction (Nikuni *et al.* 1953). A small amount of fatty acids can therefore reduce the swelling power considerably; leaching with water of 70°C. extracted 5% amylose before defatting, and 14% after (SCHOCH and FRENCH 1947).

LINDEMANN (1951) gives the following fat percentages: potato 0.20, maize 0.41, sorghum 0.32, wheat 0.24, rice 0.18. We notice that B-starches can contain fatty substances also, but that maize starch contains twice as much. This, coupled with the lower hydration, may explain the high tendency to aggregate in maize amylose. MITCHELL and ZILLMANN (1951) list the following factors which can influence the physical properties of the amylose-fatty acid complex: (1) the amount of fatty material in the complex as well as the length of the fatty acid chains; (2) the concentration and type of amylose; (3) the water-starch complex ratio. All these factors are of direct importance in respect to the swelling phenomena, displayed by various starches. Gelatinization temperature is lowered after defatting.

Monoglycerides, which have the effect of retarding staling of bread, react like fatty acids and are adsorbed by amylose (STRANDINE et al. 1951), and therefore inhibit swelling or hydration of starch granules.

The best conditions for removal of fatty acid from intact starch granules have been investigated by ALTAU (1956). Extraction is most efficient when the paracrystalline pattern has been disturbed by mechanical damage. ALTAU maintains that viscosity measurements for extracted starch indicate a strengthened surface structure, but THORNBERG (1956) could not see any change in the surface of corn starch after wet milling or methanol extraction.

There are, however, many indications that the internal structure of the starch granule has changed after fat extraction; the specific surface can also change considerably (BARHAM and CAMPBELL 1950).

For topochemical reactions the inner structure is probably more important than the surface structure.

(b) Products with artificially reduced swelling power

The swelling phenomena discussed above can be intensified in several ways for A-starches. HANSON and KATZ obtained beautifully regular "blocklet" structures by treating lintnerized wheat starch granules with $2 \, M \, Ca(NO_3)_2$. The structures were firstly made more rigid (p. 11), and were then broken up with a strong swelling agent.

"Blocklets" can often be produced when starches are very slowly heated in water, causing retrogradation, especially in the less disturbed peripheral layers (p. 22). (For the same reason wet amylose films have the tensile strength increased after heating; WOLFF et al. 1951.) When linear molecules are present, this process is practically irreversible, while retrogradation can be reversed much more easily in branched material (SCHOCH and ELDER 1953). As SCHOCH found that bread staling is mainly due to retrogradation of the branched fraction, the reversibility of this process reveals the possibility of making such bread fresh again by briefly heating to 100°C.

The tendency to retrograde is much higher in maize and wheat starch than in potato starch (MAZURS et al. 1957), and thus it is much more difficult to produce blocklets in potato starch granules by retrogradation, although they have been produced by other methods (BADENHUIZEN 1937 b).

Often a blocklet structure is found after swelling of starches in 0.5% NaOH solution. This is the result of the branched fraction becoming insoluble in the NaOH solution, in which the linear fraction is soluble (Baum and Gilbert 1956). Again the clearest effects are seen in the peripheral layers, which are least disturbed (Fig. 21).

When potato starch granules are roasted (*pyrodextrins*) chemical changes take place which increase solubility and decrease swelling power.

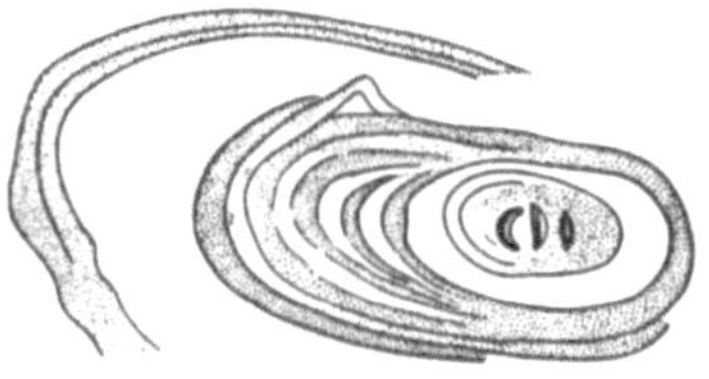

Fig. 22. Roasted potato starch granule in water showing unequal separation of layers.

The bulk of water can be removed from starch without any change taking place in the structure, but when the last three percentages are removed at 120⁰ C., some changes in the chemical composition can be demonstrated by chromatographic separation (Ulmann 1953). First large molecular complexes of the branched fraction are broken down to smaller particles, but from 150⁰ C. on dextrins appear.

These dextrins have a much more complicated structure than do the original starch molecules: their molecules are highly branched, with one out of every 6 glucose units occupying a terminal non-reducing position (Geerdes *et al.* 1957). The chains become shorter and the molecular weight decreases. The high degree of branching is the result of transglycosidation reactions, probably with the formation of new α-1,3 linkages (Christensen and Smith 1957).

The microscopical changes taking place in pyrodextrins suspended in cold water, have been systematically investigated by Badenhuizen and Katz (1938). The connections between the layers are broken, part of the substance dissolves, and the residues demonstrate a much increased swelling power in *cold* water as compared with that of the original starch. As a result the layers swell independently, often firstly in groups. The more peripherally a layer is situated, the higher its swelling pressure, with the result that layers round the centre are less well separated than peripheral layers (Fig. 22), and large granules show better separation of layers than small ones.

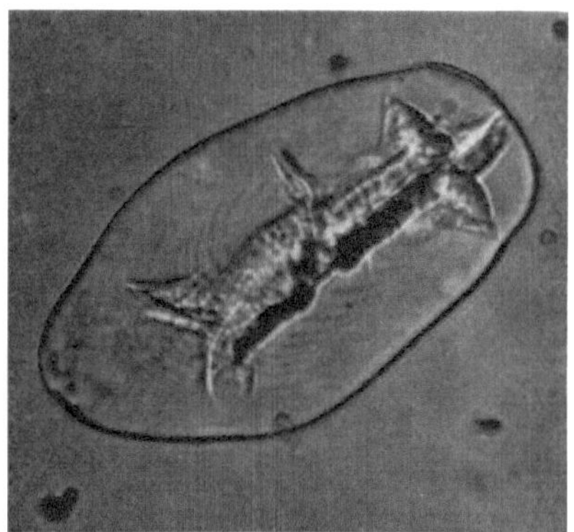

Fig. 23. Roasted potato starch granule in water, showing characteristic cracks.

In some roasted starches, like tapioca, no separation of layers takes place in water, but the granules swell in the regular way.

The isolation of layers in the described manner is independent of the X-ray diffraction pattern, as roasted wheat starch shows similar phenomena as described for potato, except that to obtain the same effect a higher roasting temperature is required.

At lower roasting temperatures large eccentric granules of potato starch show T-shaped clefts when observed in water, and gradually a series of such cracks may develop (Fig. 23). The vertical parts of the T's link up together at a later stage, splitting the granule longitudinally into two parts, which are further cleft in a transverse direction by extension of the horizontal parts of the T's. The pieces so cut off may swell independently (Fig. 24), and each even show a polarization cross, provided the swelling has not proceeded too far; the whole then gives the impression of a compound granule (cf. also Fig. 25). The parts cut off by the fissures are originally wedge-shaped, and as they swell to globular bodies, they provide one of the best illustrations of tangential swelling.

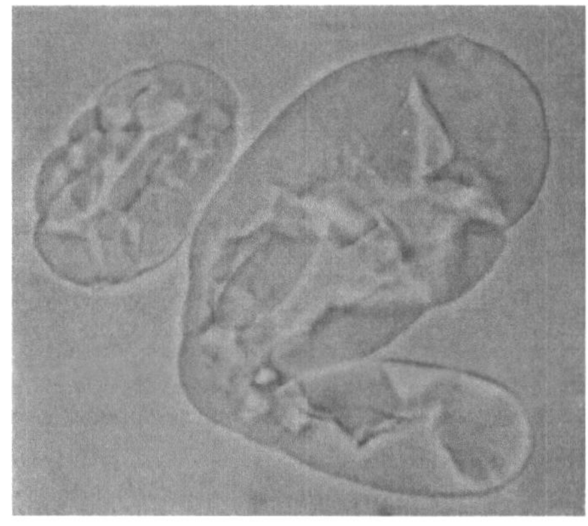

Fig. 24. Subdivision of a roasted potato starch granule into several parts by T-shaped fissures, when swelling in water.

Similar phenomena can be produced by slowly heating intact granules in water, when the swelling power is reduced by retrogradation; by the

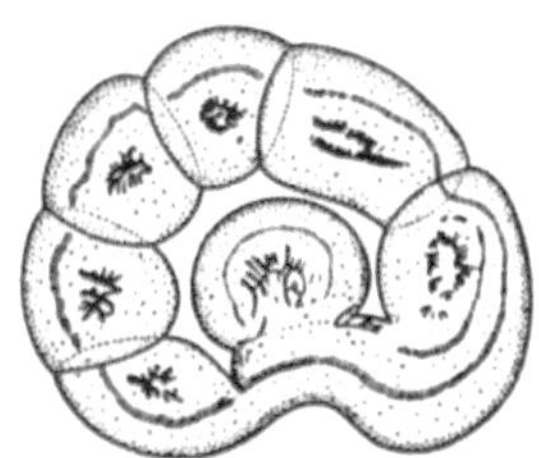

Fig. 25. Roasted maize starch granule in water, simulating a compound granule.

induction of an A-pattern by autoclaving, and by heating the starch in 30 35% alcohol. In general, by reducing the swelling power, when application of a strong swelling agent [e. g. $Ca(NO_3)_2$] can produce the phenomena at an earlier stage than can water. Granules of potato starch, heated at 140° C., will not show any perceptible changes when in contact with water, but in 2 M calcium nitrate they will produce T-fissures. This salt even produces visible changes in granules which have only been heated to a temperature of 120° C.

The granules of wrinkled pea starch have such a high primary resistance and low swelling power (p. 18), that the formation of radial clefts is a natural feature of the native granule (Fig. 26). They have occasionally been mistaken for compound granules, but they are simple as one amyloplast produces only one such starch granule (BADENHUIZEN 1955 b).

Fig. 26. Radial cracks occur naturally in wrinkled pea starch granules, which have low swelling power.

The appearance of cracks is an unfailing sign of reduced swelling power. Many of the differences in behaviour between cereal and potato-like starches find their origin in the presence or absence of fissures.

IV. Enzymic Degradation of Starch

Starch can be degraded by various enzymes, of which the amylases have been extensively studied.   Apart from amylolysis, breakdown can take place under certain conditions by phosphorolysis; amylopectin can lose its branches by the action of debranching enzymes, and several other enzymes have been found to complete the breakdown in the reaction mixture.   Whenever starch is attacked, it must be at least in a partly swollen condition: the enzymes do not act upon retrograded (strongly crystalline) starch.

It has been reported (SVESHNIKOVA 1956) that starch granules can become converted directly into fats, as found in ripening oil seeds (sunflower, flax, poppy). After fixation in an iodine solution and staining with scarlet red, the granules were seen to lose their characteristic iodine colour, while an increasingly greater part stained red, the fat being separated from the starch part by an intermediate "light zone." The writer has tried to verify this for another object, described by the Russian author, viz. the leaf stomata of *Iris* in the flowering stage. One could see a gradual decrease of starch and an increase in fatty material, but the coagulated cytoplasm in the guard cells made detailed observation very difficult. No clear pictures of transformation in single starch granules were found as described by SVESHNIKOVA.   The mechanism of this interesting conversion still is a mystery and needs further investigation.

A. Hydrolytic Enzymes

If, in the general equation given by BAILEY and FRENCH (1957):

$$G - O - R_1 + R_2\,OH \longrightarrow G - O - R_2 + R_1\,OH,$$

R_1 is a carbohydrate radical and $R_2 = H$ (G standing for glucose residue), then the equation represents irreversible hydrolysis.

1. The action of amylases

The action of amylases in vitro has been the subject of many review papers (see BADENHUIZEN 1949, BERNFELD 1951, KERR and GEHMANN 1951, HOPKINS 1954, MEYER 1952, WHELAN 1953).

Of the two types, a-amylase attacks all chains in the middle, whereas β-amylase acts at the non-reducing ends.   Their action is irreversible, notwithstanding reports to the contrary (BRESLER 1950). TALWAR *et al.* (1951) could not find any reversal of hydrolysis in their experiments, even at high pressures, and in general it is not likely that it will take place because of the loss of energy involved during hydrolytic cleavage (MACHLIS and TORREY 1956, p. 88).   Nevertheless we still find amylolysis described as an equilibrium reaction in some textbooks (MILLER 1953, p. 67).

The a—or dextrinogenic amylases are the most common in plants and animals, whereas the β—or saccharogenic amylases occur only in higher plants, as far as can be seen. Its presence has been reported for several fungi, but their saccharogenic action is probably due to a glucogenic enzyme system (FUKIMBARA and MURAMATSU 1956) (see also p. 36).

Alpha-amylases from different sources show variations in the general pattern of behaviour (Kung *et al.* 1953; Bird and Hopkins 1954 a; Sandstedt and Gates 1954), but all seem to produce α-dextrins, maltotriose, isomaltose, maltose and glucose. It is still a point of controversy whether salivary amylase can produce glucose (Pazur and Budovich 1955) or not (Giri *et al.* 1953).

Of the branched molecules both outer and inner branches (when very long) are attacked simultaneously (Lohmar 1954): linear molecules are broken down completely when enzyme concentration is high and the time of action is long (Hanrahan and Caldwell 1953).

The α-amylase from *Aspergillus oryzae* (Taka amylase) occupies a special position in that it can produce unfermentable oligosaccharides from glucose at a later stage in the reaction mixture (Pfannmuller and Noe 1952; Giri *et al.* 1953).

Beta-amylase splits off maltose from the non-reducing ends of the chains (Giri *et al.* 1953; Pazur and Sandstedt 1954). Again there is argument as to whether β-amylase can split maltotriose (Bird and Hopkins 1954 b) or not (Pazur and Sandstedt 1954). Peat *et al.* (1956) suggest that the slow action β-amylase appears to have on maltotriose in some cases may be due to the presence of traces of the so-called *D-enzyme* ($=$ Disproportioning enzyme). A mixture of β-amylase and D-enzyme is able to convert maltotriose to maltose and glucose in four hours.

Although it was earlier believed that β-amylase could degrade amylose for 100%, it was later found, with purified preparations, that degradation proceeded only up to 70% (Peat *et al.* 1949), and this has been confirmed by several authors. The interpretations however have been different. Whereas Peat *et al.* (1952) assumed the presence of an auxiliary enzyme, called *Z-enzyme*, which would be able to overcome the resistance of abnormal linkages of unknown nature in amylose, others (Hopkins and Bird 1953) thought that Z-action was simulated by traces of α-amylase. Recently the Z-enzyme idea received fresh support from Neufield and Hassid (1955). These authors, and also Baba and Kojuna (1956), demonstrated further that the supposed abnormal linkages in amylose could not be of the β-glucosidic type. The Z-enzyme has not yet been defined chemically in a satisfactory manner, and the nature of the obstacle to β-amylolysis remains unresolved for the time being.

Even synthetic amylose is not completely hydrolysed by β-amylase, although the extent of hydrolysis is greater than that for natural amylose (Neufield and Hassid 1955).

As to β-amylolysis of the branched molecules, the experiments indicate that the rate of β-amylase action gradually slows down as a 1,6-linkage (branching point) is approached. Summer and French (1956) add that although the end-groups of the β-limit-dextrin would never be any shorter than 2.5 glucose units, it might be impossible to obtain an absolute limit dextrin. This β-dextrin is multiple branched on the random pattern postulated by K. H. Meyer, like its parent amylopectin (Peat *et al.* 1956).

Alpha-amylase is stabilised by the presence of calcium ions; it is more sensitive to low pH than β-amylase, but it is less thermolabile. The behaviour of amylases in media which contain small quantities of water has been investigated by Kiermeier and Codura (1954). Below a water content of 14% there is no reaction, except when filter paper is present, which causes a different distribution of the water. Up till 43% water the reaction rate increases, to reach a constant rate at that percentage.

We know that at this level a starch suspension shows dilatancy, indicating that the thixotropic structure of the suspension is replaced by independance of the granules. Such a suspension is liquid, but immediately solidifies upon pressure (Gallay and Puddington 1943).

2. Digestibility and corrosion

Amylases act only on chains which have assumed helicoidal shape (Fig. 2). One turn of such spirals contains six glucose units, and it is possible that one whole coil needs to unite with the receptive part of a-amylase to bring about a firm union (Bird and Hopkins 1954).

Starch granules must show at least some degree of swelling before amylases can attack. It will therefore be clear that ball-milling, or heating in water, increases the susceptibility of the starch to amylase, whereas retrogradation (e. g. by freezing and thawing) causes a loss in susceptibility which is less for waxy starch than for normal maize or potato starch (Volz and Ramstad 1952).

Raw maize starch can be made much more accessible to the action of a-amylase by the addition of a cationic detergent, which permits better penetration of the enzyme into the granules. Gates and Sandstedt (1952) found that the addition of such a detergent caused better growth in chickens.

Kihara and Kawase (1953) determined digestibility by mixing starches with takadiastase at 37^0 C. and estimating the amount of reducing sugars produced after 2 weeks. Best digestibility was shown by all A-starches investigated, but bean and tapioca were comparable. Sago, potato and sweet potato were much less easily digestible.

Another method has been devised by Gates and Sandstedt (1953), who use loss of weight of the starch estimated under standard conditions, as a measure for the enzyme activity. Pancreatic amylase is most effective, then follow malt-, bacterial- and fungal-amylase (Sandstedt and Gates 1954).

The digestibility of banana starch is comparable to that of maize starch, and therefore good (Murthy and Swaminathan 1956).

The first attack on raw starch is possible only with a-amylases; β-amylase completes the breakdown when more non-reducing endgroups have become available.

Digestibility is closely linked to the swelling phenomena discussed on p. 25, and which are in turn dependent upon molecular association.

The more rigid type of starch granule develops fine cracks through which the amylase can penetrate into the interior (Fig. 20). The radially oriented molecules, which have become partly loosened from the crystalline pattern, can now be attacked sideways by a-amylase, and so the fine canals are widened into corrosion canals (Fig. 27). Wherever the enzyme reaches an amorphous layer, corrosion proceeds into the tangential direction also, and the resistent layers become more conspicuous (Fig. 28). For these reasons starches with low swelling power are more rapidly broken down by amylase than those with high swelling power, and consequently their digestibility is better.

Different wheat varieties show unequal susceptibility to excessive starch damage during swelling (BIRD 1957), and thus the structure of the individual starch granules varies in strength.

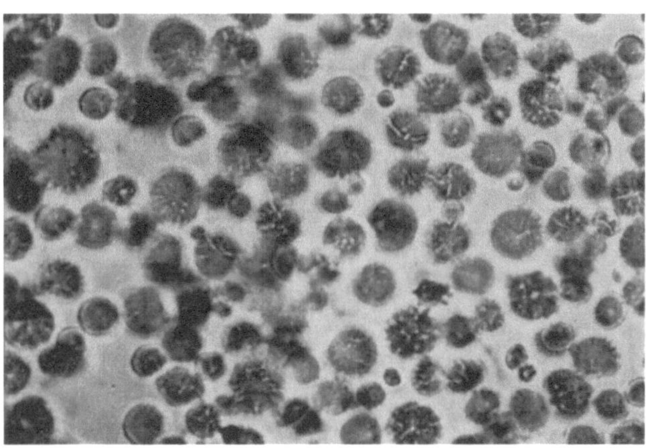

Fig. 27. Corroded starch granules of maize.

Therefore it is not surprising that there is also a difference in amyloclastic susceptibility; starch of hard wheat was found to be more rapidly hydrolysed by amylase than that of soft wheat (DARKANBAEV and KOSTYUKOVA 1956). The same authors found an increase in hydrolyzability during maturation; again, we know that during the process of maturation the first flexible transparent wheat starch granules gradually become more rigid. In the meantime SANDSTEDT (1955) reports that immature wheat starch granules are more susceptible, so one would like this point more fully investigated.

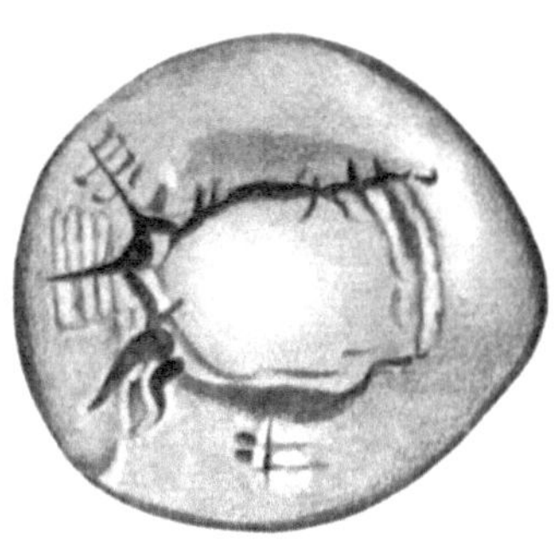

Fig. 28. Corroded starch granule of rye. (After Fig. 15 *E b* in SANDSTEDT 1955.)

Electron pictures of corrosion canals in maize starch have been published by NIKUNI and WHISTLER (1957). The ragged tooth-like shape of these canals is of course due to the sideways penetration of amylase into the amorphous layers. The central cavity is a consequence of swelling, followed by enzyme action (Fig. 7).

Similarly, cross sections made through corroded barley starch granules, show them to be in a swollen condition (Fig. 29). The two lenticular halves separate and corrosion at the narrow side weakens their connection (BADENHUIZEN 1949). Similar sec-

Fig. 29. Cross section through a corroded starch granule of barley.

tions have been illustrated by SANDSTEDT (1955, 1957), who also gave excellent descriptions and photographs of the corrosion process in wheat starch granules. However, with reference to some cases the writer would like to offer an explanation which differs from that given by SANDSTEDT.

On p. 15 it was shown that layers may swell in groups, due to their different swelling power. In the large granules of rye starch this may lead to the production of a conspicuous "amorphous" layer, which is easily attacked by amylase, and then a core separates from a ring (Fig. 28). We are not justified in calling the core a "more resistant centre island," surrounded by a "susceptible ring" representing a period in the growth of the starch granule in which the growth pattern "was changed," even if we know that this may happen in other starches (p. 56). Sandstedt states further that the "resistant island" would have the same structure as the immature granules, which he reports to be less resistant against the action of amylases than are mature granules (see above).

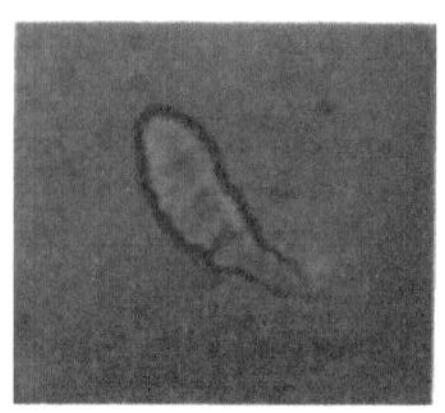

Fig. 30. Corroded starch granule of potato (phase contrast).

It would also be better not to describe the picture the corrosion canals offer with the conspicuous parts of the layers visible inside, as a series of blocks: it has nothing to do with the "blocklet" structures described on p. 25. In fact, there are no "blocks" at all (cf. Fig. 20).

These examples have been quoted in some detail to show the pitfalls of speculations, which are moreover unnecessarily complicated. There are a few fundamental features in the swelling of starch granules which are sufficient to explain most of the phenomena observed.

As corrosion proceeds swelling power decreases. For barly starch it was found that the gelatinization temperature was raised from 54° C., before, to 64° C. after, corrosion had started.

Corrosion starting from cracks is also found in a number of C-starches. Starches like potato, do not produce cracks and are corroded from the outside; potato starch granules become reduced to ragged narrow bodies. as the enzyme penetrates a little further into the amorphous layers (Fig. 30).

The phenomena described above can be imitated artificially by boiling starches in methanol to which a little sulphuric acid has been added. (Observe the suspension under the microscope, and add a drop of strong sulphuric acid to the edge of the coverslip.) Potato starch granules simply become smaller until they vanish, while maize and wheat granules form typical corrosion canals before they fall to pieces and are dissolved.

Bacteria and fungi may cause irregular pitting when in contact with potato starch granules. A pitting action, without layers becoming visible in wheat starch, characterises taka diastase (Sandstedt 1955).

According to Horning and Petrie (1928) mitochondria would play an important rôle in the corrosion of starch granules during the germination of maize, wheat and barley. Although they could also be formed in the endosperm, most of them were secreted by the scutellum, and the authors assumed that these migrated to the endosperm, passing cell walls on their way. In the endosperm they were seen to cluster around the starch granules, hydrolysing them by excreting amylase, after which they disappeared. If this were true, it would be an interesting proof for the presence of amylase in mitochondria. This is almost impossible to demonstrate for isolated mitochondria, as these lose their soluble enzymes during the process of isolation (STAFFORD 1951). KREUTZER (1956) repeated the experiments, using germinating waxy maize, and staining with Janus Green B, Nadi reagent, or haematoxylin. Although the scutellar tissue had a high concentration of mitochondria, she could not find evidence for a migration into the endosperm, and it appears most unlikely that they should be able to travel through cell walls. Three to five days after germination corrosion had started, and mitochondria could be observed along the cell walls, or generally dispersed in the endosperm cells. Nowhere were they found accumulated around starch granules. Their rôle in connection with corrosion remains at least doubtful.

ASPINALL *et al.* (1955) showed that starch from malted barley contained more of the linear fraction than would normally be found in native barley starch. This increase in amylose during germination appears to be the rule, as it was also found for germinating wheat, rice, and various other plants (FUKUI and NIKUNI 1956). In the case of barley it was concluded that during malting the free branches of amylopectin became shorter, while the amylose remained practically unaltered.

This deviation from typical amylolysis must be due to the organized structure of the starch granule, in which the relatively few linear molecules are closely and entirely associated with neighbouring chains, whereas the branched molecules are more accessible (pp. 11 and 26). We have here a good example of the difference between enzyme action *in vivo* and *in vitro* (PREECE 1957).

During the process blue value and iodine affinity increase, swelling power decreases, and gelatination temperature becomes higher, which points to increased retrogradation of the non-swollen residues.

All these phenomena could, however, also be explained by the assumption that the presence of debranching enzymes would cause the branched molecules to approach a more linear shape.

Such *debranching enzymes* hydrolyse the 1,6-glucosidic linkage and they therefore can catalyse the scission of the branches in the amylopectin molecules. They have been isolated from different sources and been given various names: isoamylase or amylosynthease from yeast (MARUO and KOBAYASHI 1951), amylo-1,6-glucosidase from muscles (CORI and LARNER 1951) or amylose isomerase (PETROVA 1952). R-enzyme from potato and *Vicia faba* (HOBSON *et al.* 1951).

Isoamylase and *R-enzyme* are probably one and the same enzyme. Waxy maize starch stains blue with iodine after its action, and in general the enzyme causes a rise in the Blue Value as the 1,6-bonds are hydrolysed. It acts upon amylopectin and β-dextrin, splitting off branches, but the action is far from complete, due to steric hindrance. R-enzyme together with β-amylase gives complete conversion of waxy maize amylopectin, but not of potato amylopectin, so that there may be resistant centres in the latter. This is possibly caused by a high degree of ramification, as the highly branched glycogen molecule is not degraded by R-enzyme + β-amylase, although it is by R-enzyme + α-amylase (PEAT *et al.* 1954, 1956). The action of R-enzyme is not reversible, as it belongs to the hydrolytic carbohydrases. The rôle it plays during germination (see above), maturation (p. 57) and the development of "blue centre spots" in waxy maize starch granules (p. 59), has still to be investigated.

The *amylo-1, 6-glucosidase* from muscles has an action different from that of the R-enzyme, in that it can not split off branches. The latter have first to be broken down by phosphorylase (P-lase, see p. 37), until the glucose unit in the 1,6-glucosidic linkage has become exposed, before amylo-1,6-glucosidase can hydrolyse this linkage. Glucose is the only fission product which can be detected (LARNER *et al.* 1951).

When R-enzyme acts upon β-dextrin the fission products are maltose and maltotriose. Their quantity (12.8%) is a measure of the proportion of free branches in amylopectin.

The combined action of phosphorylase and amylo-1, 6-glucosidase gives a mixture of glucose and glucose-1-phosphate (G-1-P). Again the proportion of free glucose to the mixture of free and esterified glucose gives an idea of the degree of branching.

With both methods the tree-like branched structure of amylopectin as proposed by K. H. MEYER was confirmed (p. 6).

Recently a debranching enzyme was described for *Saccharomyces diastaticus,* which would combine the functions of R-enzyme with that of a-1,6-glucosidase (HOPKINS and KULKA 1957).

A further type was isolated by LARNER and McNICKLE (1954, 1955) from intestinal mucosa, and called *oligo-1, 6-glucosidase.* This differs from amylo-1, 6-glucosidase in that it can hydrolyse isomaltose, panose and branched oligosaccharides. In the presence of maltase it allows for complete digestion of starch. It is probably identical to the so-called "limit-dextrinases" described earlier for barley (KNEEN and SPOERL 1948) which also were supposed to hydrolyse the 1, 6-bond, and developed during germination. Such enzymes are very important for those industries in which alcohol is produced from cereal grain.

Where fungi are used to break down starch, *glucamylases* take the place of malt-limit-dextrinases. The glucamylase of *Saccharomyces diastaticus* acted on starch, amylose, amylopctin, and β-dextrin, and was thought to be able to by-pass the 1,6-linkage. The fission product is glucose only (HOPKINS and KULKA 1957).

B. Reversible Enzymic Degradation

1. Macerans amylase

The so-called macerans amylase is excreted by *Bacillus macerans,* and produces cyclic structures of 6–8 glucose units (Schardinger-dextrins) from starch or its components. It is interesting that these dextrins give a blue-coloured complex with iodine, probably because the ring structures form tubes in which the iodine molecules can become arranged in a manner similar to that described for the helicoidal amylose molecule. A model constructed for the Schardinger-dextrin molecule with 6 glucose units has been described as a truncated cone, 12 Å in diameter and 8 Å in height (ANONYMOUS 1954).

One of the α-1,4-glucosidic linkages of Schardinger-dextrins can be hydrolysed with acid, so that an open chain is formed. From α-, β- and γ-Schardinger-dextrins respectively maltohexaose, maltoheptaose, and maltooctaose (F r e n c h *et al.* 1949, 1950), are obtained in this way.

The linear chains produced can, moreover, react with each other, so that the glucose units become redistributed over longer and shorter chains (NORBERG and FRENCH 1950).

Macerans amylase can therefore both synthesize and degrade linear chains (FRENCH *et al.* 1948), except when D-enzyme would be involved (p. 41).

2. Phosphorolysis

When the hydroxyl group of water taken up during hydrolysis is replaced by phosphate, we are dealing with phosphorolysis (or arsenolysis, when it is replaced by arsenate). The relevant enzyme, phosphorylase (P-lase), or P-enzyme, is also responsible for starch synthesis (p. 38). The degradation product is glucose-1-phosphate, and, as no loss in energy is involved, the reaction is reversible:

$$\text{Starch} + \text{phosphate} \rightleftarrows \text{G-1-P.}$$

With an excess of inorganic phosphate and without primer (p. 38), amylose is broken down completely, while amylopectin forms phosphorylase limit-dextrin. This dextrin formation has been investigated in detail by L a r n e r (1955).

The equilibrium is pH-dependent; phosphorolysis is promoted when the pH drops, and the proportion of inorganic to organic phosphate (Pi/G-1-P) increases (see also p. 38). EWART *et al.* (1954) found that this proportion was at all times favourable for phosphorolysis in living bark of *Robinia*, and therefore they did not question the possible rôle of phosphorylase in the degradation of starch in plants (p. 46). That this question is still far from being answered follows from a study by KRECH (1954). We do not know exactly what the relative importance is of phosphorylases for the breakdown of starch.

If phosphorolysis takes place there is no loss of energy, but G-1-P cannot be transported, as its permeability is extremely low (KRECH 1954). It can be used on the spot for respiration and it is then already in the glycolytic pathway. After dephosphorylation glucose is available for transport, but loss of energy is incurred, and the glucose would have to be phosphorylated again at a later stage.

The use of G-1-P for the synthesis of sucrose by means of a sucrose phosphorylase would be an obvious possibility, but such an enzyme has not been found in higher plants.

Unsatisfactory techniques are the reason why it is so difficult to assess the parts played by phosphorylase and amylases. As soon as we make an extract the enzyme activities may be entirely irrelevant in respect of the physiological conditions in the living cells, as they may become exposed to influences from which they were separated while within the cells. On the other hand, histochemical techniques are not fully enough developed to show the relative activities with certainty (Krech 1954). Moreover, activity does not need to indicate physiological function: it is easy to demonstrate the presence of phosphorylase in tissues which normally do not produce starch (p. 43) (Badenhuizen 1955 c).

Nevertheless, when amylases are present, they probably are the enzymes involved in starch breakdown, because they inhibit P-lase activity (p. 44). Where amylases are absent, P-lase can take over.

A detailed study with isotopes, as carried out by Stewart et al. (1956), who led the way in the field of protein chemistry, might possibly give an answer to what actually happens in the living cell, and in the future efforts will have to be turned more and more to this direction. It is impossible to apply directly test tube chemistry to the living material.

V. Starch Synthesis

It is interesting to note that polyglucosans, containing a high percentage of α-1,6-linkages, can be obtained by heating glucose monohydrate in the presence of cation exchangers and water (O'Colla and Lee 1956). In the following discussion the adjective "synthetic" will only be used in connection with enzymic processes.

A. Synthesis *in vitro*

Good reviews have been written by Bourne (1951), Barker and Bourne (1953), Hassid (1954), Stacey (1954), and others, so that only as much will be indicated here as is necessary to appreciate the difficulties encountered in living material.

1. Phosphorylases or P-enzymes
(a) General

These enzymes produce linear molecules (synthetic amylose) from G-1-P in the presence of a starter molecule or primer, which (for potato phosphorylase) must contain at least three glucose units linked together with α-1,4-glucosidic bonds (French and Wild 1953). The synthetic activity of phosphorylase consists of the addition of glucose units to the primer end-group, which is the non-reducing one, as substitution at the reducing end-group does not change priming ability. The function of the primer is to provide non-reducing terminal units as foundations for the chain lengthening process. Compared with the general formula on p. 30 polymer synthesis is the transfer of monomers to the acceptor.

A rise in pH from 5 to 7 causes the ratio Pi/G-1-P (p. 37) to decrease from 10.8 to 3.1, with a consequent shift of the equilibrium: starch $+$ Pi $\rightleftharpoons$ G-1-P, to the left.

The formation of a glucosidic bond requires energy, and this is provided by the C-O-P linkage of the Cori-ester.

P-lases from higher plants, for instance potato, *Phaseolus radiatus* (RAM and GIRI 1952, GIRI 1957) or Manihot (MURTHY *et al.* 1957) appear to have closely related properties. The optimum pH is at about 6, and the primers can be short linear molecules. For muscle phosphorylase the primer needs to be a large branched molecule, although also in this case only linear molecules are produced.

Synthetic amylose always has a lower D. P. than the natural product, but chain length can be improved by lowering the temperature of the reaction (HUSEMANN 1954).

(b) Primers

For potato P-lase good priming action is obtained with oligosaccharides containing four or more glucose units, the optimum primer chain length lying around 20 units. Maltotriose has a priming activity which is only 9% of that of maltotetraose (WHELAN and BAILEY 1954).

The kind and amount of primer to a large extent determine the nature of the synthetic amylose.

ONODERA *et al.* (1953) found different characteristic curves when iodine colour was plotted against time during the reaction, using glycogen, potato amylose, potato amylopectin, potato starch, synthetic starch, or malto-pentaose as primers. It is important for our understanding of the development of a starch granule, that a high concentration of primer end-groups gives a final product which has relatively short chains, while a high concentration of G-1-P or a low primer concentration induces the production of long chains (STACEY 1954). When the chains become very long their insolubility brings the reaction to a standstill (CORI 1956; also cf. p. 58).

Various P-lases may have different primer-specificity, and this factor may also influence the chemical composition of starch. Whereas soluble starch was found to be the least effective primer for P-lase from starchy or waxy maize, it was the most effective primer for the P-lase from sugary maize (TANAKA 1954).

(c) Inhibitors

A number of salts have been reported, by various authors, as inhibitors to P-enzymes, while the most diverse substances can do the same (see also for example NAKAMURA 1954). Some have been used to inhibit the enzyme specifically: $CuSO_4$ (PEAT *et al.* 1957) and phloridzin (CORI *et al.* 1943) being the most important. Glucose is a competitive inhibitor for muscle P-lase (CORI *et al.* 1943), but not for potato P-lase. It is further of great importance that amylase can inhibit plant P-lase activity directly, not only β-amylase (PORTER 1950), but probably also α-amylase (RAM and GIRI 1952, ONO 1956).

Chlorogenic acid, which is known to be concentrated in the peel of potato, effectively inhibits the P-lase (SCHWIMMER 1957). Cationic surface-active agents possibly interfere with the enzyme-substrate union and so inactivate the enzyme; full activity can be restored by means of chelating agents, which remove the toxic metallic ions from the enzyme proteins (FREDERICK 1957).

The activity of the P-lase is measured by the release of phosphate during its synthesizing action. It is quite possible that a substance has no influence on the rate of the reaction as measured by the Pi set free, but this does not mean that there is no side effect. Such a situation has been brought to light by Schwimmer and Weston (1956). Glucose and free phosphate did not influence the actual potato P-lase activity, but they had an influence on amylose formation, which was suppressed. The result was that the linear chains became shorter, and this is a point of considerable interest for further studies *in vivo* (p. 59).

2. Branching enzymes or Q-enzymes

(a) General

The Q-enzymes are responsible for the production of branched molecules from linear chains. They catalyse the direct conversion of some of the 1,4-links into 1,6-links. This process is irreversible and does not need the presence of free phosphate.

When in the general equation on p. 30, R_1 and R_2 both represent carbohydrate radicals, the equation stands for a reaction called *transglycosidation*. We saw this type of reaction when the formation of branched molecules in pyrodextrins was discussed (p. 28).

Enzymes, which bring about transglycosidation are called *transglycosidases*. They cause migration of glucosidic bonds, by splitting one, and using the energy, associated with it, to form another link.

Larner (1953) found that during the action of Q-enzyme upon chains containing 1,4-linked ^{14}C-labelled glucose units, radio-active 1,6-linked glucose units were formed. Thus the enzyme catalyzes the direct conversion of some of the 1,4-links into 1,6-links. Q-enzyme is not an amylase, nor a phosphorylase, but it is a transglycosidase with irreversible action.

Q-enzymes are much less thermostable than P-enzymes. The optimum temperature for potato Q-enzyme is 24° C. (Ram and Giri 1952), for tapioca Q-enzyme 31° C. (Murthy *et al.* 1957), whilst potato Q-enzyme becomes labile at 30° C. (Gilbert and Patrick 1952). In contrast, P-enzyme from *Phaseolus radiatus* has an optimal temperature of 55° C., and inactivation begins at 65° C. (Ram and Giri 1952).

As in the case of synthetic amylose, natural amylopectin has a much higher average molecular weight than the Q-made branched polysaccharide (Nussenbaum and Hassid 1952, Larner 1953). Higher molecular weights can only be expected when P-enzyme acts at the same time.

(b) Substrates consisting of linear molecules

As Q-enzymes are responsible for the production of amylopectin, their action must be preceded by that of P-enzymes. When the substrate consists of linear molecules, the length of these chains has an influence on Q-activity. Nussenbaum and Hassid (1952) found that chains of D. P. 23–42 were not converted, but that chains of D. P. 116 were, so that the minimum chain length on which potato Q-enzyme can act, should lie in between these values. This value is therefore higher than that given by Peat *et al.* (1953), who stated that the minimum chain length for rapid action was 40 glucose units. Below these values action is much slower.

Q-enzyme from *Oscillatoria* seems to be able to branch linear oligosaccharides with a very low D. P. (Frederick 1955). There must be at least 6 glucose units to make branching possible, as maltopentaose could not serve as substrate for the Q-enzyme. In maltohexaose and maltoheptulose branching took place only at the 5th glucose unit, so that this one could be spatially different from the same residue in maltopentaose. These results were explained by assuming a helicoidal structure having 6 glucose units per turn (cf. p. 32).

Conclusions about the action of Q-enzyme on oligosaccharides of low D. P. are, however, uncertain, as another transglycosidase, called D-enzyme, may interfere with the results.

(c) The action of D-enzyme

This enzyme catalyzes redistribution reactions among linear maltosaccharides. It has been found in potatoes, where it is closely associated with Q-enzyme (Peat *et al.* 1956, 1957). The natural substrates for D-enzyme are the maltodextrins, starting from maltotriose upwards. When it acts, for example, on maltotriose, dextrins containing up to 6 glucose units are formed. During the process two or more glucose units are transferred from a maltodextrin substrate to a suitable acceptor, D-glucose being the most active acceptor. The result is an equilibrated mixture of maltodextrins and glucose, in which no maltose is found. Only α-1, 4-links are formed.

Maltopentaose and higher dextrins are "disproportionated" to chain products, which are sufficiently long to form red iodine complexes, and which perhaps may be long enough for Q-enzyme to act upon. This explains why Q-enzyme was found to act upon maltodextrins with D. P. 3–5, whilst producing substances which stain brown-red with iodine, and sugars with an R_F-value higher or lower than that of the substrate.

Even the slow action of β-amylase on maltotriose might be due to the presence of a trace of D-enzyme.

(d) Activators for Q-enzymes

It is possible that D-activity is also responsible for an activation of Q-enzyme by the addition of maltosaccharides of low D. P., as found by Barker *et al.* (1951, 1953). Amylopectin and amylose are also activators, especially when they are broken down by mild acid hydrolysis, but maltotetraose and maltopentaose were found to be better activators.

Q-action is much more rapid on branched molecules than on linear molecules.

In general P-enzyme action is accelerated by linear molecules, and Q-enzyme action by branched molecules, so that in P/Q mixtures we are dealing with an autocatalytic process.

(e) Q - e n z y m e s   f r o m   d i f f e r e n t   s o u r c e s

Apparently there may be slight differences in the action of Q-enzymes from various sources, as Tanaka (1955, 1956) found that the free branches of amylopectin, made with maize Q-enzyme, were longer than those of amylopectin made with potato Q-enzyme. These differences could also be due to variations in the prehistory of the Q-substrates (P-enzyme and its primers, D-enzyme, see above).

(f) M i x t u r e s   o f   P -   a n d   Q - e n z y m e s

When Q-enzyme reacts with amylose, there is a progressive increase with time in the degree of branching, so that, when the reaction is stopped at different intervals, intermediate fractions can be isolated (Nussenbaum and Hassid 1952, Bebbington *et al.* 1952).

When polysaccharides were prepared with different Q/P ratios, the following changes occurred in respect of the end product with increasing quantity of Q-enzyme : solubility increased, the blue value decreased, and the properties changed from the amylose type to the amylopectin type (Foster *et al.* 1956). Certain fractions were dissimilar to any fractions ever isolated from natural starch.

Similar results were obtained by Frederick (1957) and Barker *et al.* (1950). In all cases a range of polysaccharides was formed, showing variations in molecular structure, but the synthetic product was never a mixture of amylose and amylopectin as is found in natural starch. This is one of the fundamental problems in the biochemistry of starch, which is discussed on p. 58. Frederick (1957) comes to the conclusion that the activity of Q-enzyme is the deciding factor in the type of polysaccharide synthesis by P/Q-mixtures.

B. Synthesis *in vivo*

We now have to consider in how far the *in vitro* results can be applied to what is happening in living tissues. As the phosphorylases are generally held responsible for starch synthesis, these enzymes will be considered first.

1. Phosphorylases

(a) L o c a l i s a t i o n   i n   t h e   c e l l

A good method of demonstrating phosphorylase histochemically, is to incubate sections from destarched tissues in a buffered solution of G-1-P, adjusted to pH 6 with citric acid crystals. After some time the sections are stained with iodine solution, and they are then studied under the microscope. Commercial G-1-P always contains sufficient dextrins to act as primer. Concentrations varying from 0.2% to 3% are used for incubation. The reaction is more rapid at higher temperature. For pea root tips Dyar (1950) found starch formation at 30° C. in 1 ½ hours, at 40° C. in 10 minutes. At room temperature it may take several hours.

In general starch formation under these conditions is found to take

place in the plastids, so that the plastids must contain phosphorylase.
A positive reaction was found in the chloroplasts of leaves (Shaw 1954,
Madison 1956), but also in the plastids in the yellow parts of variegated
leaves (Uno 1956), although there are exceptions. The writer found a
strong positive reaction in variegated *Coleus*, but there was no starch
formation in variegated ivy. The epidermis of Orchids and *Commelinaceae* forms particularly good material. The large leucoplasts around
the nucleus in *Tradescantia* epidermis cells all form starch after incubation
with G-1-P. In the nucleus itself the writer could never find starch accumulation, nor could Paech and Krech (1953), Krech (1954) or Duvick
(1953), although a positive reaction had been reported for the nucleus by
Dyar (1950).

Negative reactions have been reported for chloroplasts (Stocking 1952),
the starch appearing in the form of fine granules in the cytoplasm. This
phenomenon is, however, the result of damage done to the cell. G-1-P is
toxic and probably penetrates only after destruction of the semipermeability has begun. It also has an adverse influence on chloroplast structure,
which is a very labile one (Shaw 1954). For these reasons semipermeability
is often destroyed by the application of narcotics, but we do not know, in
such cases, what happens to the plastids. It is better to let the tissue die
very gradually. Permeability then increases without much damage being
done to the plastids, which now show a positive test (Paech and Krech
1953). When the plastids also become damaged, the soluble phosphorylase
diffuses out, and starch is formed around the plastids. It takes 24–28 hours
before penetration of G-1-P into cells can be demonstrated by an increase
in respiration (Krech 1954).

Duvick (1953) found phosphorylase in plastids and proplastids in the
cells of developing maize endosperm.

Demonstration of the presence of phosphorylase does not mean that
starch would have been produced under natural conditions, as was shown
for certain cells in maize endosperm (Duvick 1953) and certain tissues of
Scilla ovatifolia seedlings (Badenhuizen 1955 c). In such cases the phosphorylase has no physiological function. The originators of the phosphorylase test, Yin and Sun (1949), came to the conclusion that the site
of normal starch formation coincides with that of phosphorylase activity,
although they indicated a number of "minor" discrepancies. Badenhuizen
(1955 c) found that the test could be either positive or negative in starchcontaining tissues. The test is therefore a very artificial one, and does not
tell us much about the physiological condition of the cell (see further
pp. 46 and 58).

There may be various limiting factors, for example: lack of substrate,
and lack of primers, or the medium may be unsuitable for P-lase activity
(interference by inhibitors).

When leaves are comminuted, the vacuolar sap is mixed with the cytoplasm, and many substances are extracted which could inhibit P-lase
activity. When we consider this together with the lability of the chloroplasts, it is all the more remarkable that some investigators (Ono 1956,

Ono and Konagamitsu 1956, Sisakyan 1954) were able to obtain positive tests in isolated chloroplasts. Others never succeeded in doing this (Krech 1954, Stocking 1952), and when chloroplasts are stabilized by a coating of precipitated cytoplasm, after isolation in carbowax 4000 (Clendenning et al. 1956), and the greater part of the P-lase has become insoluble (Stocking 1956), this is no proof of its presence inside the isolated chloroplasts.

(b) Factors influencing the phosphorylase test

A number of plants do not naturally produce starch; we may call them "sugar plants." The most important inhibitors in such cases are amylases. When the amylases are inhibited by adding $^1/_{5000}$ M $HgCl_2$ to the G-1-P solution, the presence of phosphorylase can be demonstrated by the ensuing starch formation (Ono 1956). NaF ($^1/_{50}$ M) can be added simultaneously to inhibit phosphatases (Porter 1953), which also sligthly interfere with phosphorylase activity (Ono 1956, Badenhuizen et al. 1958).

Although on several occasions a relationship between acid phosphatase activity and starch production has been suspected (an inverse relationship seems to exist for example in radish roots, Ono and Konagamitsu 1957), no such relationship could be found in other plants, or in various tissues of the same plant (Badenhuizen 1955 c, Ono and Konagamitsu 1957).

Because addition of $HgCl_2$ and NaF does not bring about starch production in *Allium* during incubation with G-1-P, it is thought that no phosphorylase is present in onions (Krech 1954, Ono 1956). In the meanwhile onion juice does inhibit P-lase activity in other plants. Ono found that this inhibition could be removed by the addition of $HgCl_2$, so that it could be attributed to the presence of amylases. On the other hand Dyar (1950) reported that the inhibitor survived autoclaving at 15 lb. for 15 mins., and that it could be removed by dialysis. Here is a problem that deserves further chemical investigation.

Sugar beets were found to contain P-lase inhibitors, the activity of which showed seasonal variations and which probably were saponins (Obata et al. 1955, 1957).

Inhibitors may be localised in such a way in the cell, that they never come into contact with the P-lase, unless the tissues are minced. Leaves of *Saxifraga crassifolia* and *Tropaeolum majus* produce starch in the normal way, but their extracts are powerful P-lase inhibitors (Krech 1954), and we have seen on p. 39 that all kinds of substances may inactivate P-lase activity.

Even in one plant adjacent tissues may behave in a different way, as was first described for maize by Rhoades and Carvalho (1944). In the leaves of maize (and also *Sorghum*) the parenchyma sheath around the vascular bundles contains large chloroplasts, which produce starch, whereas the adjacent mesophyll cells contain small chloroplasts, in which no starch can be detected. When infiltrated with sugar solutions the mesophyll also produces starch.

A very similar situation was found in *Cynodon dactylon* by BADEN-HUIZEN *et al.* (1958), who could demonstrate that the lack of starch production in the mesophyll cells was due to the presence of amylases. *Cynodon* shows the characteristic feature of occasionally producing shoots in which the chloroplasts have degenerated. In such shoots the large chloroplasts of the parenchyma sheath become reduced to small colourless plastids, which lose the capacity of producing starch when incubated in G-1-P solution: the enzymic system has been destroyed in the defective plastid.

Of great interest is the fact discovered by SEYBOLD (1940), that plants lacking chlorophyll b do not produce assimilatory starch. Examples quoted by him are Brown-, Red- and Blue-Green algae, Diatoms, and *Vaucheria* of the Green algae, to which can be added *Neottia* (WEBER 1920) and a mutant of *Arabidopsis* (RÖBBELEN 1957) amongst the higher plants. SEYBOLD also mentions that in the onion chlorophyll b has become inhibited, because of the high chlorophyll a/b ratio (7–9, against the usual 3). The a/b ratio was also high in some yellow varieties, but when it was low, (2, in *Pellionia*) large starch granules were formed.

It will be worthwhile investigating the relation between chlorophyll b and phosphorylase (cf. p. 50).

Next the presence or absence of primers has to be considered. As most incubation experiments are done with G-1-P containing a certain amount of dextrins as impurifications, lack of primer could never be responsible for a negative test in this case.

DYAR (1950) found that even starved pea root tips did not need primers for the production of starch in G-1-P solution. Thus the primer might be firmly bound inside the cell, or there was a peculiar type of phosphorylase present, which did not need priming. The latter possibility was accepted for Lima beans by GREEN and STUMPF (1942), but rejected by NAKAMURA (1953). Phosphorylase from *Phaseolus radiatus* needs priming in the same way as potato phosphorylase does (GIRI 1957) (see p. 39).

No primer carbohydrate could be detected in pea roots by means of paper chromatography, but still the phosporylase, isolated from them, reacted strongly with synthetic (dextrin-free) G-1-P, so that this fraction must contain primers (BADENHUIZEN *et al.* 1957). From DYAR's work it would follow that these primers can diffuse out from frozen sections in water. They could be slowly removed from the inert protein fractions by washing; at the same time a fraction of pea phosphorylase could be isolated by fractionated precipitation, which needed a primer (MELAMED 1957). All these observations indicate that in plants like peas and beans, carbohydrate primers may be firmly adsorbed to proteins, and consequently they would be longer available after starvation than in other plants. This may also explain why pea roots can withstand starvation in distilled water much better than, for instance, maize roots, from which primer activity disappears quickly under such conditions.

The point of interest is that pea cells may always have primers available, even in starved condition. This does not necessarily apply to other plants, and comparative studies will be necessary. It is likely that primer molecules can be synthesized rapidly. Enzymes have been found in Indian pulses and in rat liver, which catalyse the synthesis of α-1, 4 linked oligo-saccharides from maltose. Such enzymes cannot be of the D-type, as maltose cannot function as a donor for D-enzyme, although it is a weak acceptor (Peat *et al.* 1956); they may be more related to Monod's amylo-maltase (Monod and Torriani 1948), catalysing the reaction:

$$n \text{ maltose} \longrightarrow n \text{ glucose} + (\text{glucose})_n.$$

In any case, the oligosaccharides produced can act as primers, and can be further modified by disproportioning enzymes.

The latter may also be responsible for those cases where the phosphory-lase test produces violet or red-staining starch (cf. Badenhuizen 1955 c. Hori 1955). When dextrins of sufficient length have been formed by P-lase, D-enzyme can transfer parts to glucose as acceptor (p. 41). This would result in the production of maltodextrins whose chains are not long enough to stain blue with iodine solution (Porter and Rees 1954).

(c) Is phosphorylase always involved in starch synthesis?

The answer to this question is obviously: no, since the enzyme may be present, while inhibited in its activity.

For the rest the above-mentioned results suggest the role of P-lase in the formation of assimilation starch as well as reserve starch. Plant organs, rich in starch, are usually rich in P-lase. When new starch layers are de-posited upon the old starch granules in potato tubers. the process starts at the far end of the tuber (Badenhuizen and Dutton 1956). which is also reported to contain most of the P-lase activity (Bolotina and Petrova 1957. Hori 1955).

The deposition of new starch is dependent upon the carbohydrate supply. It was further found that not all potatoes become active after a plant had as-similated $^{14}CO_2$, so that the transport of sugars to those tubers which were in-active, must have been blocked. This phenomenon may have been the conse-quence of faulty nutrition, as no special attention had been paid to this aspect. Lack of balanced supply of N and K has been reported as influencing accumu-lation and transport of carbohydrates in potato (Tagawa 1956). Almost all vege-tables and leaves, except the onion, show more or less P-lase activity, and little amylase activity.

On the other hand it must be mentioned that there are indications that the phosphorylase reaction may not always be involved in the synthesis of starch in plants. The ratio Pi/G-1-P for example was consistently un-favourable to the production of starch during the time that starch accumu-lated in the living bark of the locust tree .(Ewart *et al.* 1954) (p. 37). For that reason the authors considered that in this case starch formation was not catalyzed by phosphorylase, and that the results of incubation tests with G-1-P were entirely artificial.

It has been known for a long time that sucrose is the main sugar of transport in plants. In phloem juice only sucrose is found, and no sugar phosphates (Wanner 1952. Ziegler 1956), although sometimes related sugars, like raffinose and stachyose, can also be present (ZIMMERMANN 1957). After photosynthesis in $^{14}CO_2$, most of the activity is in the sucrose (BADEN-HUIZEN and DUTTON 1956, NELSON and GORHAM 1957, RADWAN 1957). We do have an idea of how sucrose phosphate can be synthesized (GANGULI and HASSID 1957). Sucrose is probably formed by hydrolysis of its phosphate by phosphatase. This can be done effectively in the vascular sheath and the phloem, which show high acid phosphatase activity (WANNER 1952, BADENHUIZEN 1955 c). The passage into the phloem and the transport appear to be related to active metabolic processes (NELSON and GORHAM 1957, ZIEGLER 1956). Sucrose contains more energy than free hexoses, and therefore it has definite advantages as the sugar of transport.

During feeding experiments with radioactive sugars it must be remembered that little of them goes into the starch, whereas after photosynthesis with $^{14}CO_2$ large amounts of ^{14}C are incorporated (VITTORIO et al. 1954). Low activity of the starch after feeding leaves with ^{14}C-uniformly-labelled glucose was confirmed by RADWAN (1957). The sucrose activity was always higher than that of the starch, the activity of G-1-P was lower throughout. It is possible that the G-1-P was rapidly converted to starch all the time. But RADWAN found further that sucrose activity started to decrease before that of the starch, and again the possible conclusion is that the P-lase mechanism may not be the only system involved in starch synthesis.

EWART et al. (1954) had suggested that the action of an amylosucrase might be responsible for the conversion of sucrose into starch, according to the equation:

$$n \text{ sucrose} \longrightarrow (\text{glucose})_n + n \text{ fructose}.$$

Such an amylosucrase is known from *Neisseria perflava* (a microbe from the human throat) (HEHRE 1949), but has not been demonstrated in higher plants. Still some such system may be found in the future and it could operate in alternation with the P-lase system, depending on the general physiological condition of the plant.

PORTER and MAY (1955) stress the fact that fructose and glucose, whether supplied free or combined as sucrose, are equally available for starch synthesis. There is a pool, consisting of equilibrated phosphate derivatives of glucose and fructose, to which sucrose contributes, and for that reason the authors think that the occurrence of an amylosucrase system is unlikely.

The whole system in the leaf is probably very complicated, as temperature, light intensity, pH, various ions, and oxygen tension are known to play a role.

In storage organs the processes involved in starch synthesis may be more simple, but they have not yet been clarified.

It is of interest that although leaves may show daily variations in activity of phosphorylase (MADISON 1956, MARRÉ and FELICI 1952), no such periodicity of the enzyme could be found in young potato tubers (BADEN-HUIZEN and MALKIN 1955), whilst sucrose concentration can show a definite periodicity, for example in the phloem of trees (ZIEGLER 1956). Moreover, when leaves are fed with sucrose solutions, their phosphorylase activity increases with sucrose concentration (Marré and Felici 1952). Thus there may be some relationship between phosphorylase and sucrose in leaves, which would be dependent upon photosynthesis, whereas in storage organs there would always be sufficient phosphorylase for synthesis.

The exact relations have not yet been found, but just as new metabolic pathways are found by-passing the Krebs cycle, so it is likely that the plant may use several possibilities for synthesis and breakdown of starch.

2. Genetical aspects of chemical composition

Genes regulate the chemical composition, the shape of the molecule, the molecular association and the crystalline pattern of the starch granule. We do not know anything, however, about the intermediate stages, which bring about this effect. Such studies in this wide field are fundamental. The most we know to date is that a particular mutant gene in maize (sh_2) causes an increase in sucrose content at the expense of the starch, so that we can say that it blocks the utilization of sucrose (Laughnan 1953). From what has been said on p. 19 it follows that it is of paramount importance to find out where the block occurs.

Some work has been done on smooth and wrinkled peas (Kellenbarger et al. 1951), from which it is learnt that high starch content is dominant over low, and low amylose content is dominant over high.

With maize interesting results have been obtained. The experiment of Parnell on the segregation of the waxy gene is classic, and shows that plants which are heterozygous for the waxy gene ($Wxwx$), produce pollen grains of which 50% will stain red with iodine solution, whilst 50% will stain blue.

The first chemico-genetic work of importance was done by Cameron (1947), who studied the effect of different gene combinations on the chemical composition of maize starch. He was followed by Kramer and Whistler (1949), Dvonch et al. (1951), Dunn et al. (1953), and Whistler et al. (1957).

The following is a composite picture from these publications. The genes studied were su (sugary, in chromosome 4), su_2 (in chromosome 6), su_x (found to be identical with su_2), su^{am} (amylaceous sugary), du (dull, in chromosome 10), and wx (waxy), and the aim was to produce a high-amylose maize starch. The genotype $SuDuWx$ gives normal starch-containing kernels, the recessive du and su genes (except su^{am}) tend to decrease starch content, but to increase amylose percentage. Of the three genes su, su_2, and du, gene su_2 is most effective in increasing amylose percentage (although this is not evident from recent figures given by Pfahler et al. 1957), and du is least efficient. When all three are combined together, maize is produced containing 9% starch of 77% amylose content. In the homozygous condition su increases amylose percentage to 50%, in combination with du it brings this figure to 65%.

The inverse relationship between starch- and amylose content was a great drawback that was recently overcome through the development of a new maize variety which has 60% starch, and 56% amylose (Deather-

AGE *et al.* 1954). This was made possible by the discovery of a gene *ha* (in chromosome 5, "*h*" standing for "high", and "*a*" for "amylose"), which changes wrinkled endosperms of low starch content to full, glassy phenotypes (KRAMER *et al.* 1956). Professor WHISTLER kindly sent the writer starch samples of the genotypes *haha* and *hasu*, which contained about 65% amylose each. Gene *ha* is now called *ae* (amylose extender) [1].

It is interesting that su_2 alone, or in combination with *du*, and the genotypes $su^{am}du$, *aeae*, and *aesu*, are now known to give a B-spectrum, while other combinations contribute to a graded series of X-ray diffraction patterns between the A- and B-spectra.

If we accept the statement that loss of birefringence is the most reliable end point to measure gelatinization temperature, then we find that the influence of various genes on this end point is extremely remarkable. For normal maize, *wx*, *du* and *duae* the end point is about 70^0 C, with amylose percentage ranging from 0–60; *ae* brings the end point up to about 90^0 C., su_2 reduces it to about 55^0, $aesu_2$ gives and end point of 83.5^0 C. (compare Table 3 on p. 18).

We have already concluded that there is no direct relation between gelatinization temperature, percentage of amylose, X-ray spectrum and average branch length (p. 19). A gene like su_2, influences conditions in the amyloplast in such a way that the starch molecules are arranged in a crystalline pattern similar to that of potato, but more labile in structure.

Thus the only way to a better knowledge of the fine structure of the starch granule, is a study of the plastids, including their physiology, colloidal structure, and genetic behaviour. This is much more difficult than the study of the starch granule itself, and has little to do with the behaviour of the molecules, as studied so far in the chemical laboratory. The starch granule is a reflection of the structure of the plastids. Only when we can reproduce the conditions under which they are formed, will we be able to bring the problems to their final solution.

VI. Growth of the Starch Granule
A. Plastids

Very young plastids may look like mitochondria, but they have their own individuality, in accordance with the hereditary features they show.

The similarity of many proplastids to mitochondria is the reason why transitional stages between mitochondria and plastids have been described (BADENHUIZEN 1939, O'BRIEN 1951, DUVICK 1955) as particles with a swollen head and a tail. According to SCHWARZ (1931) these are not transitional stages at all. With the electron microscope mitochondria and proplastids can nowadays be easily distinguished and the problem of duality is considered by most workers as having been settled (cf. FREY-WYSSLING 1955).

Proplastids, as distinguished from mitochondria, are already present in meristems. In the scutellum and the endosperm of the caryopsis they

---

[1] KRAMMER *et al.*, Agron. J. 1958. **50**. 207. 229.

are typically arranged close to the nucleus, while later they become dispersed in the cytoplasm.

The structure of chloroplasts has been discussed so often in recent reviews, that there is no need for detailed repetition in the present one. The proplastids have a core (STRUGGER's "primary granum"), from which lamellae develop, and in these the grana are formed. Blocks in pigment formation can be related to certain blocks in the submicroscopical structures (VON WETTSTEIN 1955). The *xantha* mutation of barley, for instance, lacks chlorophyll b, and has about 15% of the normal quantity of chlorophyll a (HOLMGREN 1956). Its chloroplasts show a core and lamellae, but the formation of grana is blocked. A precursor develops in the form of a great number of osmiophilic grains, and at the same time a high amount of chlorophyll b is present. When these grains degenerate, chlorophyll b disappears. It would be interesting to see if a similar scheme could be found in other plants lacking chlorophyll b (p. 45), and how this is related to phosphorylase production.

One also would like to know what the structural pecularities are of the chloroplasts (or green amyloplasts) which produce large reserve starch granules in the stem of *Pellionia daveauana,* or what are the changes, brought about in the chloroplasts of yellowing tobacco leaves (p. 16).

BARTELS (1955) made a cytological study of leucoplasts. He found that they contained grana, which were disc-shaped, but as a rule optically masked. They became visible after swelling of the stroma, staining with Rhodamin B, illumination, or fixation. The grana in the plastids of rhizomes and roots of *Iris* were visible, because these plastids are chromoplasts containing carotinoids. The proplastids contained only one granum (probably comparable to the core of prochloroplasts), and the older the plastids, the more grana they contained. Here there is a remarkable similarity to the development of chloroplasts, which must await further study with the electron microscope.

The presence of starch granules may inhibit the division of plastids (STUART 1948), but not necessarily of cells, as BRADBURY (1953) found for potato tubers. Potatoes are probably unique in this respect (PLAISTED 1957).

B. Development of Starch Granules in Plastids

1. General

In chloroplasts the starch granules are formed in the stroma, between the lamellae. The grana of *Isoetes* plastids are centres around which starch can be deposited, and which later occupy the hilum of the starch granules (STUART 1948).

Proplastids can also form starch, and this is, in fact, one of the features that distinguish them from mitochondria. In cereals, such as wheat, rye and barley, small starch granules appear when the large ones are half grown; from this period onwards the fruit starts maturing. These small granules have different properties and should not be confused with the immature lenticular granules; they may originate in proplastids.

Proplastids and plastids therefore must contain the enzyme systems, which are responsible for the production of starch.

It would be important if we could produce starch granules *in vitro,* so that it would become possible to study their development under selected conditions. Up till now this has not yet been possible.

The next best method is to study the development of starch granules in tissue cultures. Potato tuber tissue would be most suitable for this purpose. Methods for its culture have been worked out (STEWARD and CAPLIN 1951), and cell division is not hindered by the presence of starch granules. Unfortunately, the authors made no observations on starch granule development in their tissue cultures.

In the case of maize endosperm only sweet corn responded to artificial media. None of the waxy or starchy varieties could be induced to grow (STRAUS and LA RUE 1954). Whenever starch granules appear in maize endosperm cultures, they are extremely small (cf. photograph in STRAUS 1954).

Other endosperms may be more suitable. NORSTOG (1956) described the first case of starch storage in endosperm growing *in vitro,* using rye-grass. The tissue was very similar to that in the seed, and produced starch granules after 5 days, which were about as big as the nucleus. With 8% sucrose added the tissue could be kept alive for 90 days.

Attempts to grow large starch granules in sections incubated with G-1-P have failed (HESS 1955).

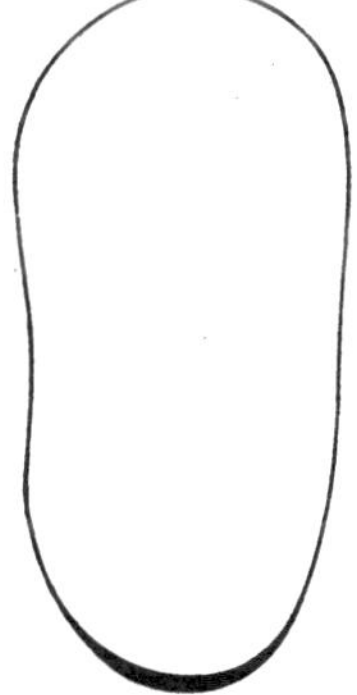

Fig. 31. Isolated amyloplast of an eccentric starch granule. (After A. MEYER 1895.)

ARTHUR MEYER (1895) showed that fundamentally the amyloplast continues to surround the starch granule during its growth (Fig. 31), except perhaps in some pathological cases. Exceptions were later reported from the cotyledons of beans and peas, and from the scutellum, where plastids failed to reappear after digestion. O'BRIEN (1951) suggests that this might be the rule for all starches showing corrosion canals, whereas in cases of peripheral corrosion the amyloplast would persist. The development of new plastids from proplastids can be stimulated by a fresh supply of carbohydrates after the old plastids have become utilized. THORNBURG (1956) states that the plastids in maize endosperm can become ruptured, which, of course, is different from utilization. It is difficult to find out in how far such ruptures are caused by the pretreatment, necessary for electron studies (p. 13). McGIVERN (1957) reports, for example, that prolonged staining with iodine solution may cause rupture of plastids and release of starch granules; exposure to glacial acetic acid can have the same effect.

Eccentric starch granules always grow inside amyloplasts which show a local thickening at the distal end of the granule (Fig. 31). The amyloplast around concentric starch granules shows equal thickness. ARTHUR MEYER (1895), whose classical book on starch granules is still highly recommended for study, attributed these phenomena to differences in viscosity,

5*

and this might well be true. It is easy to imagine that the main body of
an amyloplast with high viscosity can remain intact and be pushed to one
side by the growing starch granule. The thicker part of the amyloplast
deposits more starch. and as a result the distal layers of the granule are
thicker than the proximal ones. This is clearly illustrated by labelling the
newly deposited starch with ^{14}C. when more activity is found on the distal

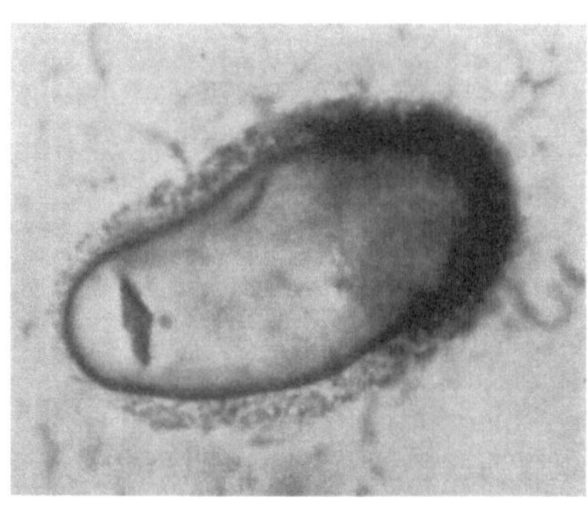

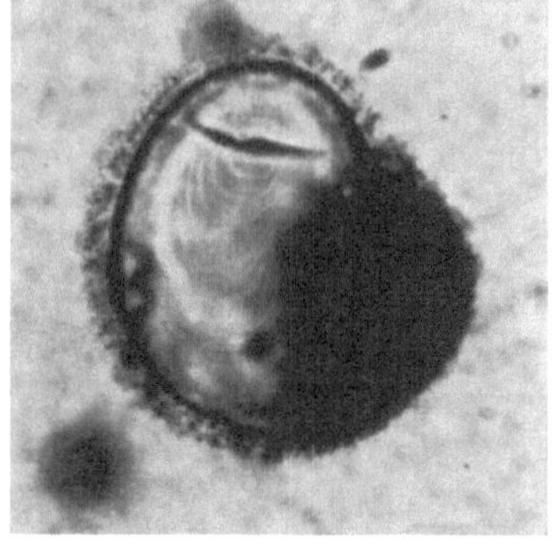

Fig. 32.   Fig. 33.

Fig. 32. Potato starch granule with newly deposited radioactive layer (Slightly roasted).
Fig. 33. Potato starch granule with strongly radioactive lateral outgrowth.

side than on the proximal side (Fig. 32). By peeling off such a layer and
showing that the starch underneath is not radioactive, Badenhuizen and

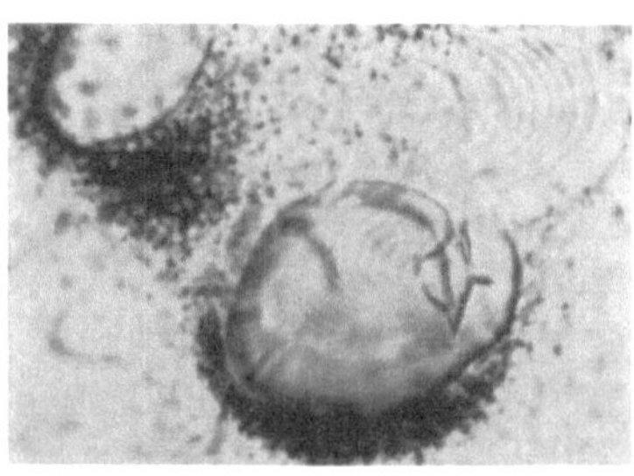

Fig. 34. Roasted radioactive potato starch
granules in water.

Dutton (1956) were able to prove beyond
doubt. that a potato starch granule grows
by apposition. When an amyloplast shifts
its position around a granule. a lateral
outgrowth is formed which is highly
radioactive (Gafin 1957; Fig. 33).

Similar studies of another B-starch
(*Oxalis latifolia*). and further mani-
pulations with active potato starch granules. confirmed this picture. although
one has to be careful with the interpre-
tation of swelling phenomena when a
redistribution of the starch substance takes place. When active starch
granules are roasted they become entirely inactive as soon as they show
separation of layers in water (p. 28). whereas one would expect some
activity to persist in the outermost layer. Intermediate stages (Fig. 34)
demonstrate that the outermost layer becomes dispersed at an early stage.
Figs. 32 and 34 show that the active material disappears first at the proximal end and lingers at the distal end (Gafin 1958).

That there are differences in viscosity of plastids might follow from
a study by McGivern (1957). Within the matrix of large mature intact
plastids in the sieve tubes of *Helianthus, Gossypium* and *Nicotiana*, the

starch granules showed Brownian movement. This movement was not seen in similar plastids in *Cucurbita*, which were also not ruptured, so that their structure must have been different.

When radioactive sucrose reaches the potato tuber, a weak activity is found to spread out evenly throughout the tuber. After the concentration has reached a certain level, starch formation begins at the tip of the tuber and proceeds towards the base, as can be followed in autoradiographs of sections. The starch granules become active very gradually, which indicates that starch may be deposited in very thin layers, depending upon the carbohydrate supply (BADENHUIZEN and DUTTON 1936).

In the amyloplast starch molecules are built up, which crystallize out. This procress is repeated only after sufficient sugars have diffused in to replace the former. When transport is blocked, no starch deposition takes place.

According to this theory, all types of starch granules should possess a layered structure (with the limitations mentioned on p. 13), which in fact they do.

The plastid, with its particular structure, regulates the deposition of starch. Water content of the plastid, density of structure, the presence of various substances, and probably other factors, all contribute to the construction of the starch granule. In the writer's opinion the abundance of water during crystallization could be a major factor in determining the type of crystalline pattern.

The foregoing pages have shown, that whatever happens in the amyloplast, it must be a very complicated process.

One further problem is the radial orientation of the molecules in the starch granule. When we mix 1 g. starch (or its components) with 5 ml. 40% formol in a Petri dish, and we keep the lid closed, after three days a clear gel forms. Allowing the formol to evaporate causes a turbidity to appear on the surface, brought about by beautifully layered spherocrystals (Fig. 43). Examination with the polarization microscope reveals that birefringence of these spherites is much lower than that of starch granules, and that their molecules are arranged in tangential fashion. This may be caused by the same forces that make rod-shaped virus particles assume a position parallel to the edge of a droplet, when this is drying out (cf. Fig. 33 on p. 71 in LURIA 1953).

All long molecules, which can form H-bonds, are able to produce spherulites. A necessary condition is the formation of nuclei which can grow when numerous favourably situated chain molecules are present in the immediate neighbourhood. When this happens in a medium, where surface tensions cannot interfere (as they do during the process of drying), then the arrangement of the molecules is a radial one. We find this for instance, in artificial spherites of inulin, which develop in *Dahlia*-tuber tissue when treated with alcohol. We also find a radial arrangement in natural starch granules, indicating that this process has nothing to do with the structure of the plastid, but that it is a purely mechanical one.

2. Changes during development

(a) Changes in shape

Starch deposition is not always as regular as has been described in the

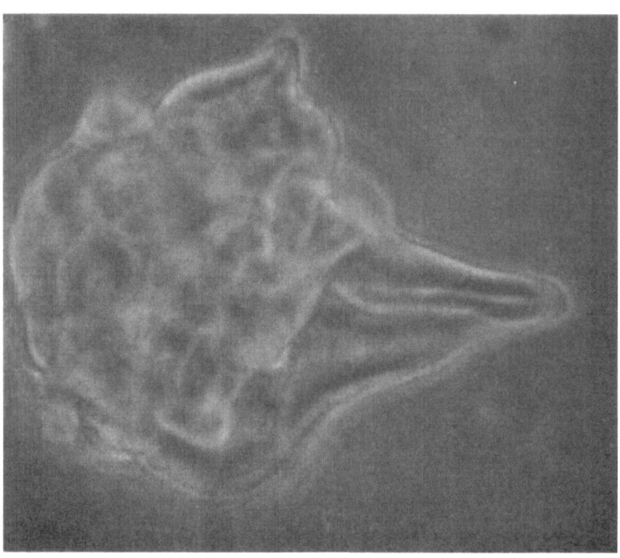

Fig. 35. Starch granule with irregular layers from a bulb-scale of *Scilla ovatifolia* (Phase contrast).

previous pages. Important changes may take place in the amyloplast, causing a radical change in the mode of deposition. Sometimes the granule becomes covered with an irregular mass, showing many projections (Fig. 35, Fig. 36 top left).

Such irregularities can often be shown to arise from the fusion of a great many nuclei, both in natural starch granules and in formol spherites. The polarization cross disappears in such structures, although they are still birefringent. We imagine that this could happen in a viscous medium with a high concentration of raw material.

Another interesting feature is the fusion of several granules of different sizes by common layers, as is for instance, found in the older bulb scales of *Scilla ovatifolia* (Fig. 36). Such structures are a clear illustration of apposition, and their history can easily be read. What cannot be read, and has to be investigated, is the behaviour of the amyloplasts from which these granules arise. It it possible that several plastids fuse and then continue to produce common layers for their respective

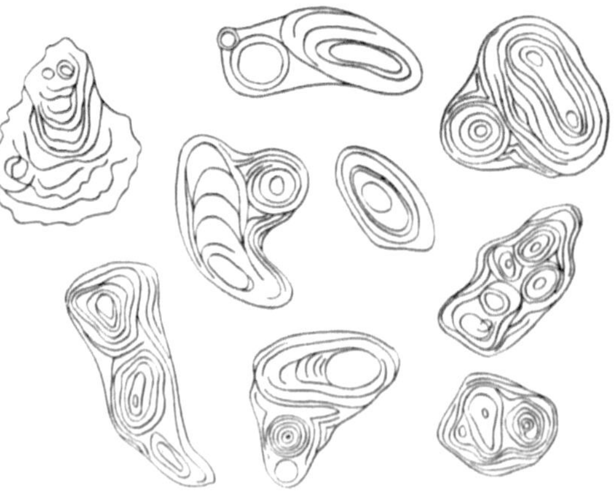

Fig. 36. Semi-compound starch granules from the bulb-scales of *Scilla ovatifolia*. In the middle a normal simple granule for comparison. Granules lintnerized.

starch granules? In that case we would be dealing with aggregates, rather than with compound granules.

A very similar picture is given by the *aesu*-genotype of maize (p. 49) the *aeae* genotype giving more regularly formed granules, Fig. 5, which are also more easily lintnerized, and show a normal polarization cross. In Fig. 37 the same feature as mentioned for *Scilla* (Fig. 35), is shown:

a number of granules have produced numerous outgrowths, which represent as many nuclei of crystallization. Others appear to be single, but in practically all of them birefringence is absent. These "grape-like" structures are very similar to an abnormal type of starch granule, which is sometimes found in potatoes, stored in the refrigerator

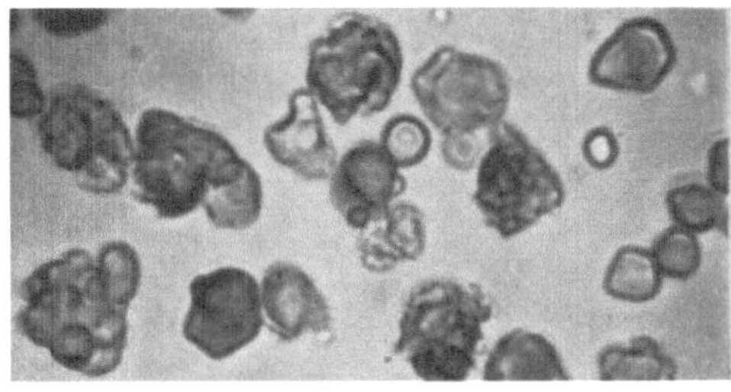

Fig. 37. Starch granules from the endosperm of maize with *aesu*-genotype.

and afterwards kept at room temperature (Fig. 38). As the amount of sugar increases during the treatment, the abnormal starch granules in potato

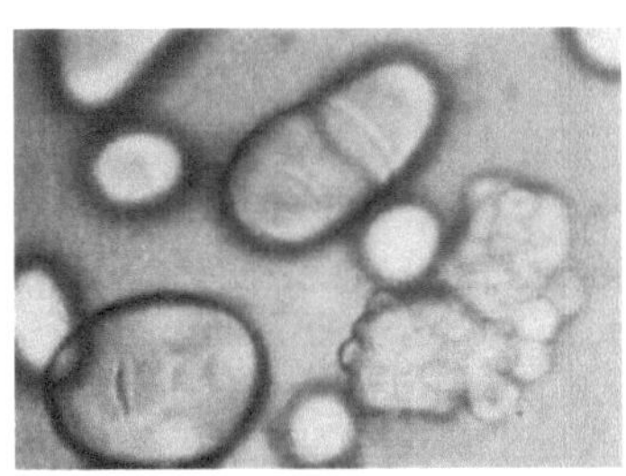

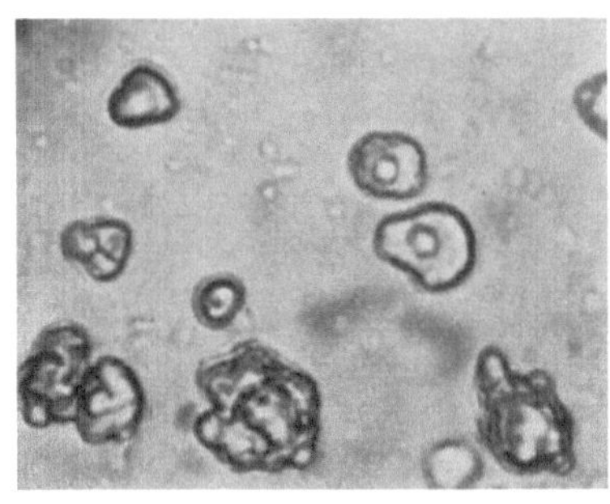

Fig. 38.   Fig. 39.

Fig. 38. Two "grape-like" starch granules (right) and two normal ones (left) from a cold-stored potato.

Fig. 39. Starch granules of *aesu*-maize, showing nuclei.

may be the result of many nuclei originating simultaneously and fusing afterwards. Both from the theoretical and practical points of view it is

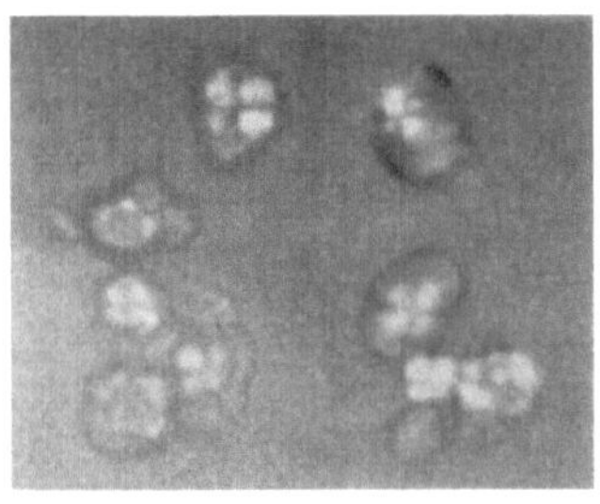

Fig. 40. Starch granules of *aesu*-maize between partly crossed nicols.

of interest to know the exact conditions under which the "grape" structures are formed.

After lintnerization especially it becomes clear that many of the *aesu* granules show one or more nuclei (Fig. 39). Between crossed nicols only these nuclei show birefringence and a regular polarization cross, whilst the substance around it appears to be optically isotropic (Fig. 40: in order to take this picture polarisator and analysator were not set at exact right angles to each other, to show the contours of the whole granule).

Stages in the development, which have to be investigated more fully, are to be found in the cells of the endosperm 18 days after pollination.

In Fig. 41 various types of granules have been brought together, which may all occur in one cell.

Birefringence has been indicated by a cross, except for the smallest granules. Evidently starch granules are produced in a regular way at an

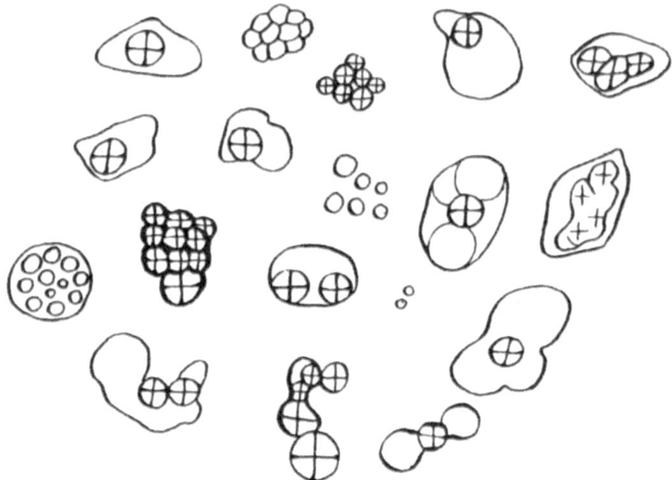

Fig. 41. Stages in the development of *aesu*-maize starch granules, as seen in the endosperm 18 days after pollination.

early stage, but later an optically isotropic mass of starch is deposited around one or more of the original granules, leading to structures which are highly irregular in shape. The original granules may become entirely masked, but are sometimes still discernable by their birefringence. In this way aggregates are built up. As the mature starch gives a sharp B-spectrum the isotropic part must be of a crystalline nature. In a number of cases a wavy layered structure could be detected in it. The immature granules have a much higher swelling power than the mature ones.

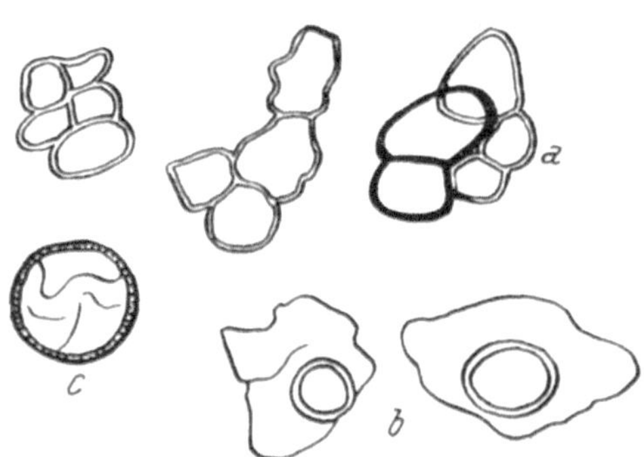

Fig. 42. Swollen starch granules of *aesu*-maize. *a* Compound granules. *b* and *c* stages during swelling of granules with a nucleus (in stage *b* granules are still bright under phase contrast, in stage *c* they look black.)

When *aesu* starch granules are heated in water the aggregates retain their composite structure (Fig. 42). The nucleus is seen swelling to a sac inside the isotropic mass, but at a later stage of swelling no trace of the nucleus is left (Fig. 42 *b* and *c*).

Nuclei, surrounded by material deposited at a later stage, are also found in *aeae*-granules, but the secondary deposit is of a much more regular nature than that of *aesu*-granules, so that the finished granule shows the ordinary polarization cross. The nuclei stand out well against the surrounding layer when lintnerized granules are treated with chloral hydrate, and are then observed under phase contrast. Treatment with saliva also makes them clearly visible.

It is interesting that similar phenomena can be produced in formol-spherites (Fig. 43). Apart from the free regular spherite with layers (and often "blocklets"!), there are others which have become surrounded by irregular masses. These peripheral deposits are birefringent.

On one occasion the writer found one or more starch granules surrounded by an isotropic mass of starch, after incubation of waxy maize endosperm in G-1-P solution. The phenomenon was strikingly similar to what is described above for *aesu* starch and can evidently be produced without the help of plastids.

(b) Changes in structure

In normal starch the most frequent size of the granules increases during development, for normal and waxy, for instance, from 3 μ to 9 μ during the first three weeks after pollination. At the same time water content of the maize fruit drops from 87% to 10% (WOLF *et al.* 1948). The immature starch granules in rice, barley, wheat and maize (and probably in grasses in general) are more smooth, transparent, flexible, stainable, and less birefringent, than the mature granules. As less and less water is available, retrogradation increases. These stages can be influenced by environmental conditions.

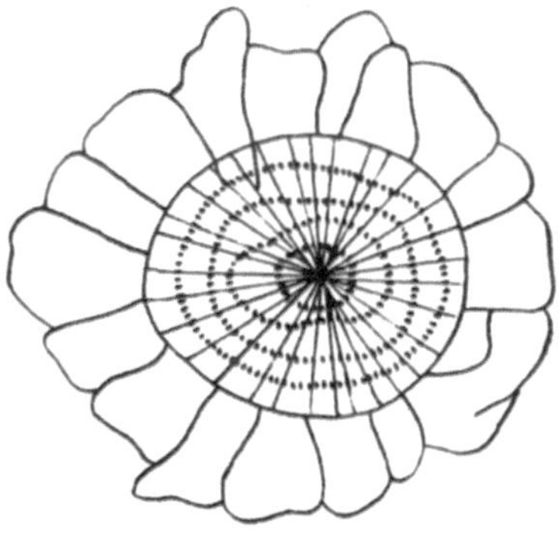
Fig. 43.   Formol spherite from waxy maize.

Parallel with the changes goes an increase in viscosity of the paste, and a decrease in swelling power, or a rise in gelatinization temperature (HARRIS and JESPERSON 1946, MAYWALD *et al.* 1955, SCHOCH and MAYWALD 1956, UEDA *et al.* 1955) (see also p. 33).

Various authors (WOLF *et al.* 1948, MAYWALD *et al.* 1955, FUKUI and NIKUNI 1956) report an increase of the blue value during ripening, but whether this is due to an increase in the actual amount of linear component, or to an increase in chain length, cannot be deduced from these experiments (see also p. 35) and care must be exercised with interpretations.

Immature stages are more susceptible to enzyme action than mature ones, although the pattern of action is the same (SANDSTEDT 1955).

For potatoes STEPANENKO and AFANASEVA (1957) maintain that fractionation of starch from young tubers is more difficult than that from old tubers, and they ascribe that to an increase in molecular weight of the branched molecules, accompanied by less branching.

All these statements remain fairly vague, but deserve thorough investigation, as we shall see from the following.

3. The presence and distribution of linear molecules in natural starch granules

(a) Possible factors involved

The elucidation of the fact that there are linear molecules at all in natural starch granules, offers one of the most challenging problems in the biochemistry of starch. We have seen that mixtures of P- and Q-enzymes will produce only branched molecules (p. 42).

One can assume that the 25% amylose, found in ordinary starch, represents a residue which escaped the action of the Q-enzyme. The question, of course, is why this should be so.

Frederick (1956) studied a mutant in *Oscillatoria princeps* which produced a starch with mainly linear molecules (instead of the glycogen-like carbohydrate found in the wild type). The mutant however contained the branching enzyme just as the normal type did, and no inhibitor could be found.

When waxy endosperm is incubated with G-1-P solution, blue-staining starch is formed, although under natural conditions there is a complete conversion of linear to branched molecules (Badenhuizen 1955 c, Fuwa 1957 a), which is further evidence for the production of long chains in a medium with high G-1-P concentration (p. 39).

One could expect that the P/Q ratio would vary for different species and varieties, but Fuwa (1957 b) found the same ratio in starchy and waxy maize. It looks as if a study of the enzymes does not add to the solution of the problem. They are all present, but there must be factors regulating their activity.

The late Prof. G. T. Cori (1956) made the important observation that when the concentration of glycogen as primer for muscle P-lase is low, the enzyme forms linear molecules, which are so long that they become insoluble. Such molecules could not act as a substrate for Q-enzyme and would remain unbranched.

On p. 39 the importance of primer concentration for the structure of the linear chains produced by P-lase was discussed.

As a working hypothesis the writer would like to suggest that during the formation of the normal starch granule primer concentration is low. As a result a number of linear molecules would be able to escape branching, in the manner indicated by Cori (1956), before the autocatalytic process of branching (p. 41) would reach sufficient speed to convert the remainder of the linear molecules to amylopectin. A high concentration of primer molecules would produce a great number of shorter chains, which would not have a chance of escaping branching, thus leading to waxy types of starch. To prove this, a survey will have to be made of the available primers in various plants, which produce starch having a different amylose percentage.

This theory is based upon the individuality of the plastids, which has been receiving ample support in recent years (Badenhuizen 1956). One characteristic feature of the plastid could be the nature and the concentration of its primers. This does not exclude the possibility of more than one type of plastid occurring inside one cell (p. 60). The fact that starch granule size increases in polyploids indicates a possible influence of the nucleus. In tetraploid valerian. granules had a maximum size of 18 μ. in octaploid plants this increased to 30 μ (SANYAL and WALLIS 1957).

The same fundamental process has to take place during the deposition of each new layer. with the result that each layer will have the same chemical constitution. and this is in accordance with the observations (see p. 11. also BADENHUIZEN 1955 a).

BAKER and WHELAN (1951) assumed a gradual increase in amylopectin content in centrifugal direction. In the light of our working hypothesis this would mean a gradual increase of primer concentration in the plastid. which is *a priori* unlikely for the formation of ordinary starch.

The theory fits in well with the conclusion arrived at earlier (BADEN-HUIZEN 1955 a), that in the common type of starch there is a regular distribution of linear molecules throughout the granule.

The concentration of G-1-P is not likely to have much influence, as it has always been found to be low. Another factor may be the presence of phosphate. which tends to suppress amylose formation in keeping the chains short (SCHWIMMER and WESTON 1956). It would be interesting to see. for instance. whether the plastids in young developing endosperm of waxy maize would show a higher phosphate content than those in starchy maize.

(b) Waxy starch granules with "blue centre spots"

A great number of waxy starch granules possess a nucleus, which stains blue with iodine solution, and which will be called "centre spot" for the sake of brevity. Such starch granules are the notable exception to the general rule of equal distribution of linear molecules. and they pose the most difficult problem of all. They are mentioned in A. MEYER's book. but the first systematic description for waxy maize was given by LAMPE (1931), who found that with distance from the centre of the endosperm starch granules had increasingly larger centre spots. until these would occupy the whole granule. Generally these centre spots are small in waxy maize granules. When normal maize starch is lintnerized for one month. the granules still stain blue with iodine. The same applies to the centre spots, and this gives an easy method for their isolation (BADENHUIZEN 1955 a).

For waxy maize starch the percentage of amylose is generally given as 1%. but DEATHERAGE *et al.* (1955) mentioned a number of varieties in which the amylose percentage varied considerably.

Would the size of the centre spot increase with amylose content?

Dr. E. ANDERSON. Director of Missouri Botanical Gardens. kindly sent me a few such varieties for investigation. They were studied after lintner-ization. when the distribution of blue-staining material becomes clearly visible.

The results are tabulated in Table 4.

From this little survey it is clear that there is no question of proportion between size of centre spot and amylose percentage. At a low percentage linear molecules can be distributed evenly throughout the granule, or they are localized in smaller or larger spots in the centre. There is, however, a clear increase in staining intensity with increasing amylose percentage.

The indications are that cereal starch granules also grow by apposition. When they are labelled with C-14 and then subjected to the action of

Table 4. *Distribution of linear fraction in waxy varieties with different amylose content*

Starch from	Amylose percentage	Size of centre spots
Argentine waxy corn	3	Absent or small
Stonor H 4 (maize)	4	Whole granule stains blue
Stonor 16 Assam (maize)	7	Absent, or of any size, until occupying entire granule
Waxy barley Spartan	7	Of any size, until occupying whole granule (Fig. 44)
McNeilly Pop (maize)	9	Granules stain evenly, but with varying intensities

amylase, the outermost layer remains active, while the degraded inside (p. 33) shows no activity (GAFIN 1957; see photographs in SANDSTEDT, 1955).

The first starch granules formed in the endosperm of waxy maize all stain brown with iodine. At the same time brown-staining granules start

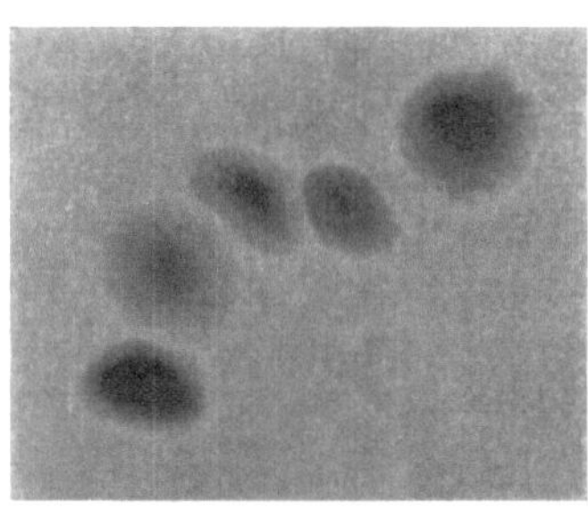

Fig. 44. Waxy barley starch granules, after lintnerization and staining with iodine.

to appear amongst the blue staining ones in the pericarp, so that we may find the two types side by side in one cell. In a later stage of development some blue-staining granules make their appearance amongst the brown-staining granules in the endosperm cells (KREUTZER 1956).

From these observations we can see that at certain periods conditions change in the cells, and that these changes become reflected in the properties of the newly formed plastids.

It may even be possible that properties of a plastid change during its activity in starch production. It could start off with the formation of normal blue-staining starch, to be followed by the deposition of branched molecules only, perhaps because of an increase in primer concentration.

These phenomena have been mentioned mainly to demonstrate the complexity of the problems confronting us in the field of starch biochemistry. Many are of a fundamental nature. There are still wide gaps in our knowledge of the starch granule, but to fill them is a very attractive task for biologist and chemist alike.

Acknowledgements

The author is grateful to the South African Council for Scientific and Industrial Research, for financial support: to Mr. J. E. Gafin, who prepared many photographs, and for his general assistance: to Mr. M. Melamed for general assistance: to Mr. H. J. Swart, for drawing Fig. 20: to Mrs. B. Pike, for making most of the drawings: to Prof. R. L. Whistler (Purdue University), for sending samples of high amylose maize: to Dr. E. Andersen (Missouri Botanical Gardens). for sending samples of waxy varieties: to Prof. H. J. Logie for placing an X-ray apparatus at our disposal: and to Miss F. D. Hancock for reading the manuscript.

Rückblick

Seit mehr als einem Jahrhundert hat das Stärkekorn Chemiker und Biologen gefesselt. Nach einer Zeit der größten Verwirrung wurde aber in den Jahren nach dem letzten Kriege Übereinstimmung erreicht über die Haupttatsachen der chemischen Zusammensetzung. und es wurde das Vorhandensein zweier Molekültypen anerkannt: das lineare und das verzweigte Molekül. beide aufgebaut aus Glukoseeinheiten. Damit war noch nichts ausgesagt über die wirkliche Größe dieser Moleküle im nativen Stärkekorn, ihre Verteilung, und ihre Bildung. Zwar wurden die Enzyme, welche die Stärkemoleküle aufbauen, entdeckt und konnte ihre Wirkung im Reagenzglas verfolgt werden. aber die Anwendung der rein chemischen Ergebnisse auf das Geschehen in der lebendigen Zelle stößt noch immer auf große Schwierigkeiten.

Das Problem der Chemiker ist. das Verfahren während der Analyse möglichst schonend zu gestalten. damit keine Artefakte entstehen. Aber auch der Mikroskopiker hat damit zu kämpfen. daß kleine Eingriffe schon bedeutende Folgen haben können. und daß demnach die Unterscheidung natürlicher und abgeleiteter Strukturen recht schwierig werden kann.

So darf es nicht wundernehmen. daß noch täglich neue Resultate ans Licht kommen. Es sollte nie vergessen werden. daß die Stärkekörner grundsätzlich b i o l o g i s c h e Objekte sind. deren Erzeugung an lokale Strukturen in der Zelle gebunden ist: die Plastiden. Daher kann man nur weiter kommen durch das Studium der Plastiden. deren leichte Verletzlichkeit jedoch die schwierigsten Probleme schafft. Interessante Aussichten werden aber von chemogenetischer Seite eröffnet.

Obgleich die Stärkekörner tote kristallinische Gebilde sind, spiegeln sie Vorgänge in den Amyloplasten wider, Vorgänge, welche teilweise beeinflußt werden von bestimmten Genen, teilweise aber auch dem Einfluß äußerer Faktoren unterworfen sind. Die Beschaffenheit des radial orientierten molekularen Gefüges der verschiedenen Stärkearten ist wahrscheinlich abhängig von der verfügbaren Menge freien Wassers während der Stärkeablagerung im Amyloplasten. und daher von den physiologischen Verhältnissen des betreffenden Organs. Diese Verhältnisse können sich während der Entwicklung ändern. und es gilt daher die Faktoren, welche die Stärkeprezipitation in den Plastiden regulieren, genauer zu untersuchen.

Die zwei synthetisierenden Enzyme, Phosphorylase (P-Enzym) und Q-Enzym, können in den meisten Gewebeextrakten nachgewiesen werden. Dennoch wird eine ganze Reihe von Stärkesorten mit variierendem Amylosegehalt gefunden. Das Vorhandensein von linearen Molekülen in nativen Stärkekörnern in genetisch bestimmten Mengen ist eines der Grundprobleme der Stärkeforschung. Die Anfangskonzentration und Art der Kohlehydrate, welche die Phosphorylasereaktion anregen ("primers") könnten hier eine Rolle spielen.

In Mischungen von P- und Q-Enzymen werden *in vitro* immer verzweigte Moleküle erzeugt. Das Maß der Verzweigung nimmt entsprechend zu, wenn die relative Konzentration des Q-Enzyms gesteigert wird. Daher können wechselnde Enzymmengen, oder die Annahme spezifischer Inhibitoren, keine Basis zur Erklärung des Amylosegehaltes bilden. Beide Enzyme sind außerdem wasserlöslich, und eine örtliche Trennung in den Amyloplasten wäre daher unwahrscheinlich. Es ist klar, daß eine Anzahl von linearen Molekülen der verzweigenden Wirkung des Q-Enzyms entgehen kann, und daß dieser Prozeß in den verschiedenen Stärkearten mehr oder weniger erfolgreich ist, am wenigsten in solchen, welche nur verzweigte Moleküle aufweisen (Klebestärke).

Es scheint weiter möglich zu sein, daß der Grad des Erfolges sich während des Wachstums des Stärkekorns ändert, so daß man Klebestärkekörner findet, deren Mitte sich mit Jodlösung blau färbt (Klebemais, Klebegerste). Die individuellen Klebestärkekörner zeigen diese Erscheinung in verschiedenem Ausmaß.

Wir wären daher gezwungen anzunehmen, daß in einzelnen Fällen der Wechsel der Kohlehydratstoffe der Zellen einer Änderung unterworfen ist, welche die „Anreger" der Phosphorylasereaktion beeinflußt; dafür spricht, daß zur Zeit der Stärkebildung im Endosperm des Klebemaises auch im Perikarp mit Jodlösung sich braunfärbende Stärkekörner zwischen den normalen finden.

Weitere Faktoren, welche in das Zustandekommen und die Verlängerung linearer Moleküle eingreifen könnten, sind die vorhandenen Mengen von anorganischem und organischem Phosphat (Glukose-1-Phosphat = G-1-P). Ein Überschuß von G-1-P produziert lediglich lineare Moleküle, aber eine niedrige Konzentration erzeugt nur wenige längere solcher Moleküle.

Diese Anschauung steht in Übereinstimmung mit der Annahme, daß die Phosphorylase das beherrschende Enzym der Stärkesynthese ist, und daß die einzelnen Stufen der Verwirkung des Rohmaterials, und G-1-P selbst, in bestimmten Strukturen lokalisiert sind.

Die Analyse des wachsenden Kartoffelstärkekorns mittels ^{14}C hat sichergestellt, daß Apposition vorherrscht, und es gibt Hinweise, daß dasselbe auch für die anderen Stärkearten gilt. Wir müssen uns daher vorstellen, daß der gleiche Grundprozeß der Kristallisation aller Stärkeschichten vorangeht. Solange sich die Struktur des Amyloplasten nicht ändert, solange wird jede Schicht unter genau denselben Bedingungen gebildet werden, und wird die Struktur aller Schichten dieselbe sein. Ob eine neue Schicht geformt wird, oder ob sie breit oder schmal wird, ist von der Zufuhr der Kohlehydrate, hauptsächlich der Saccharose, abhängig.

Es wird noch immer nach einem besonderen Enzym gesucht, das für die direkte Umwandlung von Saccharose in Stärke verantwortlich sein sollte, aber Enzyme, deren Existenz bereits bewiesen ist, könnten die Saccharose vollständig in G-1-P verwandeln (D. H. Turner und J. F. Turner, Austr. J. Biol. Sci. 1957, **10**, 302).

Die Moleküle der neuen Schicht werden teilweise in die alte Schicht eindringen, und folglich sollten die aufeinanderfolgenden Schichtgrenzen unsichtbar sein. Der Quellungsdruck nimmt aber mit der Entfernung vom Mittelpunkt zu und veranlaßt eine gegenseitige Verschiebung der Schichten, welche das Entstehen von sogenannten „amorphen Schichten" als Begrenzung der kristallinischen Schichten hervorruft. Die amorphen Schichten werden daher sekundär gebildet; sie sind breiter, wo sie G r u p p e n von Schichten begrenzen, als wo sie die e i n z e l n e n Schichten begrenzen. Wir erwarten daher keine Korrelation mit dem täglichen Wechsel von Licht und Dunkel, und tatsächlich werden Schichten auch unter konstanten Bedingungen gebildet.

Wir können zugleich eine Erklärung geben für die niedrig molekulare Amylosefraktion, welche bei vorsichtigem Auslaugen mit heißem Wasser als erste aus dem Stärkekorn diffundiert (J. M. G. Cowie und C. T. Greenwood, J. Chem. Soc. **1957**, S. 4640): sie stammt von den amorphen Schichten, in denen der molekulare Verband zerstört worden ist.

Die größeren Moleküle müssen erst aus den mizellaren Verbänden, wo sie durch Wasserstoffbrücken zusammengehalten werden, unter Steigerung der zugeführten Energie herausgelöst werden, aber die molekulare Dispersion dürfte nur teilweise gelingen.

Obgleich die Heterogenität der linearen und verzweigten Moleküle wohl plausibel gemacht wurde, ist es bisher doch nicht gelungen, die Existenz einer Übergangsfraktion eindeutig nachzuweisen. Eine leichte Verzweigung mancher Amylosemoleküle ist nicht ausgeschlossen, wird allerdings durch das Studium des sogenannten Z-Enzyms nicht bewiesen (J. M. Cowie, I. D. Fleming, C. T. Greenwood und D. J. Manners; J. Chem. Soc. **1957**, p. 4430). Es scheint, als ob eine Anzahl Moleküle die verzweigende Aktivität umgehen können, während andere Moleküle völlig verzweigt werden, als Folge eines autokatalytischen Prozesses.

Die Anziehungskräfte zwischen den Ketten werden beeinflußt von deren Länge, von vermischten Fettsäuren, verfügbarem Wasser, und dem Amylosegehalt. Ein höherer Gehalt an linearen Molekülen könnte die radiale Orientierung verzerren, was zur Abschwächung der Doppelbrechung und Verstärkung der Molkohäsion führt, allerdings unter Beibehaltung der Kristallinität. Die neuen Maisvarietäten mit hohem Amylosegehalt sind in diesem Zusammenhang sehr lehrreich. Es kommt aber noch mehr dazu, denn ein einziges Gen (su_2) kann das Gerüst so weit lockern, daß die Verkleisterungstemperatur bedeutend herabsinkt. Es scheint demnach unmöglich, einfache Regeln aufzustellen für die Beziehung zwischen Verkleisterungstemperatur, Doppelbrechung, Amylosegehalt, Röntgenspektrum und Kettenlänge, und weitere Erläuterungen der Unterschiede in der Feinstruktur der Stärkearten müssen künftigen Untersuchungen vorbehalten bleiben. Inzwischen sind sie aufs engste verbunden mit der Verdaulichkeit der Stärke.

Eine kräftigere Molkohäsion fördert das Entstehen von feinen Rissen, welche durch die Amylasen zu Korrosionskanälen erweitert werden können. Ein solcher Angriff von innen heraus macht das Stärkekorn der Verdauung besser zugänglich. Die Quellung von Stärkekörnern mit erniedrigter Quellkraft kann Veranlassung sein zur Bildung von Artefakten („Blöckchen"), welche keine präformierten Strukturen sind. Sie ließen sich auch nicht ins allgemeine Schema des Wachstums einfügen.

Die Quellung ist eine tangentiale, wobei die Schichten zusammenfließen, um die Blasenwand zu bilden. Es wird wohl heute kein Forscher mehr die Existenz einer Außenmembran mit besonderen Eigenschaften annehmen.

Es wurde versucht, in dem vorliegenden Artikel die Lücken in unserer Kenntnis und die Unvereinbarkeit mancher Ergebnisse und Behauptungen anzudeuten. Der Autor ist sich der Unvollkommenheit seiner Darstellung bewußt. Es dürfte aber doch deutlich geworden sein, wo die Probleme liegen: die Fragestellung hat sich weitgehend von der chemischen nach der b i o l o g i s c h e n Seite verschoben. Damit ist sie schwieriger, aber auch von grundsätzlicher Bedeutung geworden. Die mühsame Arbeit ist nicht beendet, sie fängt eben an. Wie es häufig in der Wissenschaft der Fall ist, kann man auch hier sagen: wir wissen viel und doch erst wenig.

References

Alsberg, C. L., 1938: Structure of the starch granule. Plant Physiol. **13**, 295.

Altau, K., 1956: The consequences of solvent extraction on the structure and behaviour of the starch granule. Dissertation Abstr. **16**, 1209.

Anderson, D. M. W., and C. T. Greenwood, 1955: Interaction of starches and branched α-1,4-glucosans with iodine; and a valve microvoltmeter for differential potentiometric titrations. J. Chem. Soc., p. 3016.

— — 1956: Carbohydrates of the roots of the Parsnip, *Pastinaca sativa*. J. Chem. Soc., p, 220.

— — and E. L. Hirst, 1955: The oxidation of starches by potassium metaperiodate. J. Chem. Soc., p. 225.

Anonymous, 1954: Starch atoms in 3-D. Chem. Eng. News **32**, 4184.

— 1957: Chem. Eng. News **35**, 61, 89.

Aspinall, G. O., E. L. Hirst, and W. McArthur, 1955: Constitution of a modified starch from malted barley. J. Chem. Soc., p. 3075.

Baba, A., and H. Kojuna, 1956: Absence of β-glucosidase-labile linkage in amylose. Chem. Abstr. **50**, 16898 h.

*Badenhuizen, N. P., 1937 a: Die Struktur des Stärkekorns. Protoplasma **28**, 293.

— 1937 b: Untersuchungen über die sogenannte Blöckchenstruktur der Stärkekörner. Protoplasma **29**, 246.

— 1938: Das Stärkekorn als chemisch einheitliches Gebilde. Rec. Trav. Bot. néerl. **35**, 559.

— 1939: Growth and corrosion of the starch grain in connection with our present knowledge of the microscopical and chemical organization. Protoplasma **33**, 440.

— 1941: Verdere onderzoekingen over het marmerverschijnsel. Meded. Proefstation voor Vorstenlandsche Tabak **89**, 75.

---

* means Reviews.

BADENHUIZEN, N. P., 1946: On the swelling of starch grains. Transact. Faraday Soc. **62** (B), 265.
*— 1949: De chemie en de biologie van het zetmeel. Oosthoek (Utrecht).
*— 1951: Structure and irreversible degradation of starch in relation to the baking process. Bakers Digest **25**, 21.
— 1953: Swollen starch grains and osmotic cells. Experientia **9**, 136.
*— 1954: The growth of starch grains in relation to genotype and environment. S. Afr. J. Sci. **51**, 41.
— 1955 a: Observations on the distribution of the linear fraction in starch granules. Cereal Chem. **32**, 286.
*— 1955 b: The structure of the starch granule. Protoplasma **45**, 315.
— 1955 c: Distribution of acid phosphatase and phosphorylase in relation to starch production. Acta Bot. Neerl. **4**, 565.
— 1956: Some aspects of cytoplasmic inheritance. S. Afr. J. Sci. **53**, 82.
— and R. W. DUTTON, 1956: Growth of ^{14}C-labelled starch granules in potato tubers, as revealed by autoradiographs. Protoplasma **47**, 156.
— and J. R. KATZ, 1938: Mikroskopische Beobachtungen an den durch Röstdextrinbildung veränderten Stärkekörnern. Z. physik. Chem. (A) **182**, 73.
— and M. I. MALKIN, 1955: Phosphorylase activity in relation to the layering of starch granules. Nature **175**, 1134.
— J. E. BARTLETT, and E. N. GUDE, 1958: Preliminary investigations on the anatomy, histochemistry and biochemistry of normal and chlorotic leaves in *Cynodon dactylon* Pers. S. Afr. J. Sci. **54**, 37.
— E. J. L. KREUTZER, and L. ZAR, 1957: Demonstration of primer-independence for pea phosphorylase. S. Afr. J. Sci. **53**, 203.
BAILEY, J. M., and D. FRENCH, 1957: The significance of multiple reactions in enzyme-polymer systems. J. Biol. Chem. **226**, 1.
BAKER, F., and W. J. WHELAN, 1950: Birefringence of amylose and amylopectin in whole structural starches. Nature **166**, 34.
— — 1951: Birefringence, iodine reactions and fine structure of waxy starches. J. Sci. Food Agric. **2**, 444.
BALLOU, C. E., and E. G. V. PERCIVAL, 1952: The structure of the sapwood starch of the maple (*Acer* spp.). J. Chem. Soc., p. 1054.
BARHAM, H. N., and C. L. CAMPBELL, 1950: Some effects of the progressive absorption and desorption of palmitic acid on the behaviour of a Sorghum grain starch. Transact. Kansas Acad. Sci. **53**, 235.
*BARKER, S. A., and E. J. BOURNE, 1953: Enzymic synthesis of polysaccharides. Quat. Rev. **7**, 56.
— A. BEBBINGTON, and E. J. BOURNE, 1951: Carbohydrate primers for Q-enzyme. Nature **168**, 834.
— — — 1953: The mode of action of the Q-enzyme of *Polytomella coeca*. J. Chem. Soc., p. 4051.
— E. J. BOURNE, S. PEAT, and I. A. WILKINSON, 1950: The use of mixtures of P- and Q-enzymes in the synthesis of starch-type polysaccharides. J. Chem. Soc., p. 3022.
BARTELS, F., 1955: Cytologische Studien an Leukoplasten unterirdischer Pflanzenorgane. Planta **45**, 426.
BAUER, A. W., 1953: The fractionation of starch. Ph. D. thesis, Princeton University.
BAUM, H. A., and G. A. GILBERT, 1954: The anaerobic dispersion of potato starch. Oxygen-sensitive bonds in starch. Chem. a. Ind., p. 489.
— — 1956: The fractionation of potato starch by centrifugation in alkali. J. Coll. Sci. **11**, 428.
— — and H. L. WOOD, 1955: The fractionation of normal and mottled wheat starch. Viscometric properties of the amylose fraction. J. Chem. Soc., p. 4047.
BEBBINGTON, A., E. J. BOURNE, and I. A. WILKINSON, 1952: The conversion of amylose into amylopectin by the Q-enzyme of *Polytomella coeca*. J. Chem. Soc., p. 246.
BECHTEL, W. G., 1951: Examination of starch dispersions with the phase microscope. Cereal Chem. **28**, 29.
BENGOUGH, W. I., P. R. STANWIX, and W. STEEDMAN, 1957: The isolation of amylose and amylopectin from barley starch. Chem. a. Ind., p. 1241.
*BERNFELD, P., 1951: Enzymes of starch degradation and synthesis. Adv. Enzymol. **12**, 379.

Bird, L. H., 1957: Role of damaged starch in the evaluation of wheat quality. Nature **180**, 815.

Bird, R., and R. H. Hopkins, 1954 a: The action of some α-amylases on amylose. Biochem. J. **56**, 86.

— — 1954 b: The mechanism of β-amylase action. II. Biochem. J. **56**, 140.

Bolotina, T. T., and A. N. Petrova, 1957: Peculiarities of enzymic transformations of carbohydrates in various parts of the potato tuber. Chem. Abstr. **51**, 5214 i.

Bottle, R. T., G. A. Gilbert, C. T. Greenwood, and K. N. Saad, 1953: Degradation of potato amylose in neutral and alkaline solution. Chem. a. Ind., p. 541.

*Bourne, E. J., 1951: Some recent developments in the academic fields of starch chemistry. Chem. a. Ind., p. 1047.

— 1952: Some recent developments in the academic fields of starch chemistry: a correction. Chem. a. Ind., p. 216.

Bresler, S. E., 1950: Synthesis of protein and starch. Chem. Abstr. **44**, 4529 h.

Bünning, E., und C. Hess, 1954: Schichtenbildung im Stärkekorn bei konstanten Außenbedingungen. Naturw. **41**, 339.

Burgeff, H., 1932: Saprophytismus und Symbiose. Fischer (Jena).

*Caldwell, C. G., 1952: Industrial starch research. Chem. Eng. News **30**, 514.

Cameron, J. W., 1947: Chemico-genetic bases for the reserve carbohydrates in maize endosperm. Genetics **32**, 459.

Campbell, W. G., J. L. Frahn, E. L. Hirst, D. F. Packman, and E. G. V. Percival, 1951: Wood starches. I. J. Chem. Soc., p. 3489.

Christensen, G. M., and F. Smith, 1957: The constitution of a wheat starch dextrin. J. Amer. Chem. Soc. **79**, 4492.

Cillie, G. G., and F. J. Joubert, 1950: Occurrence of an amylopectin in the fruit of the Granadilla (*Passiflora edulis*). J. Sci. Food Agric. **1**, 355.

Clendenning, K. A., T. E. Brown, and E. E. Walldov, 1956: Causes of increased and stabilized Hill reaction rates in polyethylene glycol solutions. Physiol. Plantarum **9**, 519.

Cori, C. F., G. T. Cori, and A. A. Green, 1943: Crystalline muscle phosphorylase. III. Kinetics. J. Biol. Chem. **151**, 39.

*Cori, G. T., 1956: Enzymatic structure analysis and molecular weight determination of polysaccharides. Makromol. Chem. **20**, 169.

— and J. Larner, 1951: Action of amylo-1,6-glucosidase and phosphorylase on glycogen and amylopectin. J. Biol. Chem. **188**, 17.

Coton, L., L. H. Lampitt, and C. H. F. Fuller, 1955: Starch structure. II. Determination of iodine absorption by amperometric titration. J. Sci. Food Agric. **6**, 660.

Cowie, J. M. G., and C. T. Greenwood, 1957 a: The effect of acid on potato starch granules. J. Chem. Soc., p. 2658.

— — 1957 b: Aqueous leaching and the fractionation of potato starch. J. Chem. Soc., p. 2862.

Cramer, F., und W. Herbst, 1952: Die Lichtabsorption von Jodkettenmolekeln. Naturw. **39**, 256.

Czaja, A. T., 1954: Mikroskopischer Nachweis von Amylose und Amylopektin am Stärkekorn sowie zweier Typen von Stärkekörnern. Planta **43**, 379.

— 1955: Colour photography in the study of starch. Photogr. u. Forschung **6**, 123.

Darkanbaev, T. B., and T. N. Kostyukova, 1956: Enzymic hydrolyzability of starch from various varieties of wheat. Chem. Abstr. **50**, 5850 f.

Deatherage, W. L., M. M. MacMasters, and C. E. Rist, 1955: A partial survey of amylose content in starch from domestic and foreign varieties of corn, wheat, and Sorghum and from some other starch-bearing plants. Transact. Amer. Assoc. Cereal Chem. **13**, 31.

— — M. L. Vineyard, and R. P. Bear, 1954: Starch of high amylose content from corn with high starch content. Cereal Chem. **31**, 50.

Dimmick, R. L., and R. C. Corley, 1950: A modified starch iodine method suitable for the study of starch and its hydrolysis products. Proc. Indiana Acad. Sci. **60**, 141.

Doremus, G. L., F. A. Crenshaw, and F. H. Thurber, 1951: Amylose content of sweet-potato starch. Cereal Chem. **28**, 308.

Dunn, G. M., H. H. Kramer, and R. L. Whistler, 1953: Gene dosage effects on corn endosperm carbohydrates. Agron. J. **45**, 101.

Duvick, D. N., 1953: Phosphorylase in the maize endosperm. Bot. Gaz. **115**, 82.

Dvonch, W., H. H. Kramer, and R. L. Whistler, 1951: Polysaccharides of high-amylose corn. Cereal Chem. **28**, 270.

Dyar, M. T., 1950: Some observations on starch synthesis in pea root tips. Amer. J. Bot. **37**, 786.

Ewart, M. H., D. Siminovitch, and D. R. Briggs, 1954: Possible enzymatic processes involved in starch-sucrose interconversions. Plant Physiol. **29**, 407.

Fischer, E. H., and W. Settele, 1953: Fractionnement chromatographique de l'amidon. Helv. **36**, 811.

Foster, A. B., P. A. Newton-Hearn, and M. Stacey, 1956: Ionophoresis of carbohydrates. Part III. Behaviour of some amylosaccharides and their reaction with borate ions. J. Chem. Soc., p. 30.

Foster, J. F., and E. P. Paschall, 1953: Effect of disaggregation of amylose on the properties of the iodine complex. J. Amer. Chem. Soc. **75**. 1181.

— and D. Zucker, 1952: Length of the amylose-iodine complex as determined by streaming dichroism. J. Phys. Chem. **56**, 170.

Frederick, J. F., 1955: Proposed molecular structure for straight and branched polymers of glucose. Physiol. Plantarum **8**, 288.

— 1956: Physico-chemical studies of the phosphorylating enzyme of *Oscillatoria princeps*. Physiol. Plantarum **9**, 446.

— 1957: Effect of surface activity and chelation phenomena on the activity of the polyglucosid-synthesizing enzymes of *Oscillatoria*. Physiol. Plantarum **10**. 844.

French, D., and G. M. Wild, 1953: Primer specificity of potato phosphorylase. J. Amer. Chem. Soc. **75**. 4490.

— D. W. Knapp, and J. H. Pazur, 1950: Studies on the Schardinger dextrins. VI. The molecular size and structure of the γ-dextrin. J. Amer. Chem. Soc. **72**, 5150.

— M. L. Levine, and J. H. Pazur, 1949: Studies on the Schardinger dextrins. II. Preparation and properties of amyloheptaose. J. Amer. Chem. Soc. **71**. 356.

— J. Pazur, M. L. Levine, and E. Norberg, 1948: Reversible action of macerans amylase. J. Amer. Chem. Soc. **70**, 3145.

Frey-Wyssling, A., 1953: Submicroscopic morphology of protoplasm. Elsevier (Amsterdam).

— 1955: Die submikroskopische Struktur des Cytoplasmas. Protoplasmatologia, vol. 2 (A 2), p. 27.

Fukimbara, T., and K. Muramatsu, 1956: The action of mold amylase on alcoholic fermentation. I. Chem. Abstr. **50**, 14866 d.

Fukui, T., and Z. Nikuni, 1956: Degradation of starch in the endosperm of rice seeds during germination. J. Biochem. **43**, 33.

Fuwa, H., 1957 a: Formation of starch in young maize kernels. Nature **179**, 159.

— 1957 b: Phosphorylase and Q-enzyme in developing maize kernels. Arch. Biochem. Biophys. **70**, 157.

Gafin, J. E., 1958: M. Sc. thesis (Dept. of Botany. University of the Witwatersrand, Johannesburg).

Gallay, W., and I. E. Puddington, 1943: The hydratation of starch below the gelatinization temperature. Canad. J. Res. **21 B**, 179 (see also p. 171).

Ganguli, N. C., and W. Z. Hassid, 1957: Synthesis of sucrose in *Impatiens holstii*. Plant Physiol. (Suppl.) **32**, XXXIV.

Gassner, G., 1955: Mikroskopische Untersuchung pflanzlicher Nahrungs- und Genußmittel. Fischer (Stuttgart).

Gates, R. L., and R. M. Sandstedt, 1952: Effect of a cationic detergent on the digestion of raw corn starch *in vitro*. Science **116**. 482.

— — 1953: A method of determining enzymatic digestion of raw starch. Cereal Chem. **30**, 413.

Geerdes, J. D., B. A. Lewis, and F. Smith, 1957: The constitution of corn starch dextrins. J. Amer. Chem. Soc. **79**, 4209.

Gilbert, G. A., and A. D. Patrick, 1952: Enzymes of the potato concerned in the synthesis of starch. Biochem. J. **51**, 181, 186.

*Giri, K. V. 1957: Enzymic synthesis of starch. Sympos. on starch and amylases, Osaka, p. 25.

— V. N. Nigam, and K. Saroja, 1953: Application of circular paper chromatography to the study of the mechanism of amylolysis. Naturw. **18**, 484.

Green, D. E., and P. K. Stumpf, 1942: Starch phosphorylase of potato. J. Biol. Chem. **142**, 355.

*Greenwood, C. T., 1956: Aspects of the physical chemistry of starch. Adv. Carbohydr. Chem. 11, 335.
— and J. S. M. Robertson, 1954: The characterization of the starch present in the seeds of the Rubber Tree, *Hevea brasiliensis*. J. Chem. Soc., p. 3769.
Hannsen, E., E. Dodt, und E. G. Niemann, 1953: Bestimmung von Korngröße, Kornoberfläche und Korngewicht bei pflanzlichen Stärken. Kolloid-Z. 130, 19.
Hanrahan, V. M., and M. L. Caldwell, 1953: A study of the action of taka amylase. J. Amer. Chem. Soc. 75, 2191.
Harris, R. H., and E. Jesperson, 1946: Factors affecting some physicochemical properties of starch. Food Research 11, 216.
*Hassid, W. Z., 1954: Biosynthesis of complex saccharides. Chemical Pathways of Metabolism. Edited by D. M. Greenberg. Academic Press Publ. (New York). Vol. 1, p. 235.
*— and E. W. Putman, 1950: Transformations of sugars in plants. Ann. Rev. Plant Physiol. 1, 109.
Hehre, E. J., 1949: Synthesis of a polysaccharide of the starch-glycogen class from sucrose by a cell-free bacterial enzyme system (amylosucrase). J. Biol. Chem. 177, 267.
Hellman, N. N., and E. H. Melvin, 1950: Surface area of starch and its role in water sorption. J. Amer. Chem. Soc. 72, 5186.
— T. F. Boesch, and E. H. Melvin, 1952: Starch granule swelling and water vapor sorption. J. Amer. Chem. Soc. 74, 348.
— B. Fairchild, and F. R. Senti, 1954: The bread staling problem. Molecular organization of starch upon aging of concentrated starch gels at various moisture levels. Cereal Chem. 31, 495.
Hess, C., 1955: Über die Rhythmik der Schichtenbildung beim Stärkekorn. Z. Bot. 43, 181.
Hess, K., H. Mahl, E. Gütter, und E. Dodt, 1955: Elektronenmikroskopische Beobachtungen an Schnittflächen von Weizenkörnern. Mikroskopie 10, 6.
Hester, E. E., A. M. Briant, and C. J. Personius, 1956: The effect of sucrose on the properties of some starches and flours. Cereal Chem. 33, 91.
Hirai, N., 1953: Swelling velocity of starch and lyophilic property. Chem. Abstr. 47, 2569 a.
Hobson, P. N., W. J. Whelan, and S. Peat, 1951: R-enzyme. J. Chem. Soc., p. 1451.
Holmgren, P., 1956: Some observations on the amount of protein and non-protein nitrogen in two pigment mutations of barley. Kungl. Lantbruks-Högskolans Ann. 22, 353.
*Hopkins, R. H., 1954: The action and properties of beta-amylase: recent developments. Wallerstein Lab. Commun. 17, 299.
— and R. Bird, 1953: The Z-enzyme in amylolysis. Nature 172, 492.
— and D. Kulka, 1957: The glucamylase and debrancher of *Saccharomyces diastaticus*. Arch. Biochem. Biophys. 69, 45.
Hori, S., 1955: Formation of starch in storage organs. Chem. Abstr. 49, 7657 i.
Horning, E. S., and A. H. K. Petrie, 1928: The enzymatic function of mitochondria in the germination of cereals. Proc. Roy. Soc. 102 B, 188.
Houven van Oordt-Hulshof, R. van der, 1957: Cell-wall structure of some Monocotyledones. Acta Bot. Neerl. 6, 420.
Huggins, M. L., 1957: Hydrogen bonding in high polymers and inclusion compounds. J. Chem. Educ. 34, 480.
Husemann, E., 1954: Über natürliche und synthetische Amylose. Die Stärke 6, 2.
Katz, J. R., und T. B. v. Itallie, 1930: Alle Stärkearten haben das gleiche Retrogradationsspektrum. Z. physik. Chem. 150 (A), 90.
Kawamura, S., and H. Fukuba, 1957: Legume starches. II. Viscosity behaviour. Chem. Abstr. 51, 15709 f.
Kellenbarger, S., V. Silveira, R. M. McCready, H. S. Owens, and J. L. Chapmans, 1951: Inheritance of starch content and amylose content of the starch in peas (*Pisum sativum*). Agron. J. 43, 337.
Kerr, R. W., and F. C. Cleveland, 1952: The structure of amyloses. J. Amer. Chem. Soc. 74, 4036.
*— and H. Gehman, 1951: The action of amylases on starch. Die Stärke 3, 271.
Kiermeier, F., und E. Codura, 1954: Über den diastatischen Stärkeabbau in lufttrockenen Substanzen. Biochem. Z. 325, 280.
Kihara, Y., and Z. Kawase, 1953: Digestibility of starch and starchy foods. Chem. Abstr. 47, 7126 i.

KITE, F. E., T. J. SCHOCH, and H. W. LEACH, 1957: Granule swelling and paste viscosity of thick-boiling starches. Bakers Digest **31**, 42.

KNEEN, E., and J. M. SPOERL, 1948: Limit-dextrinase activity of barley malts. Proc. Amer. Soc. Brew. Chem. **13**, 20.

KNYAGINICHEV, M. I., Y. R. B. BOLKHOVITINA, and T. A. MAKSAKOVA, 1955: Significance of the coating membrane of starch grains in relation to the properties of starch. Chem. Abstr. **49**, 9308 e.

KRAMER, H. H., and R. L. WHISTLER, 1949: Quantitative effects of certain genes on the amylose content of corn endosperm starch. Agron. J. **41**, 409.

— — and E. G. ANDERSON, 1956: A new gene interaction in the endosperm of maize. Agron. J. **48**, 170.

KRECH, E., 1954: Über die Phosphorylase in höheren Pflanzen. Beitr. Biol. d. Pflanzen **30**, 379.

KREUTZER, E. J. L., 1956: Unpublished results (Botany Department. University of the Witwatersrand, Johannesburg).

KUNG, J. T., V. M. HANRAHAN, and M. L. CALDWELL, 1953: A comparison of the action of several alpha amylases upon a linear fraction from corn starch. J. Amer. Chem. Soc. **75**, 5548.

LAMBERT, J. L., 1951: Linear starch reagents. Cadmium iodide-linear starch reagent. Anal. Chem. **23**, 1247.

— and S. C. RHOADS, 1956: Preparation of linear potato starch fraction for quantitative colorimetric iodimetry. Anal. Chem. **28**, 1629.

LAMPE, L., 1931: A microchemical and morphological study of the developing endosperm of maize. Bot. Gaz. **91**, 337.

LAMPITT, L. H., C. H. FULLER, and N. GOLDENBERG, 1948: The fractionation of potato starch. Part IV. The absorption spectra and colour intensities of the starch-iodine complexes. J. Soc. Chem. Ind. **67**, 97.

LANSKY, S., M. KOOI, and T. J. SCHOCH, 1949: Properties of the fractions and linear subfractions from various starches. J. Amer. Chem. Soc. **71**, 4066.

LARNER, J., 1953: The action of branching enzymes on outer chains of glycogen. J. Biol. Chem. **202**, 491.

— 1955: The action of crystalline muscle phosphorylase on outer chains of glycogen and amylopectin. J. Biol. Chem. **212**, 9.

— and C. M. McNICKLE, 1954: Action of intestinal extracts on "branched" oligosaccharides. J. Amer. Chem. Soc. **76**, 4747.

— — 1955: Gastrointestinal digestion of starch. I. The action of oligo-1, 6-glucosidase on branched saccharides. J. Biol. Chem. **215**, 723.

— B. A. ILLINGWORTH, and G. T. CORI, 1951: Enzymatic analysis of structure of branched polysaccharides. Feder. Proc. **10**, 213.

— — — and C. F. CORI, 1952: Structure of glycogens and amylopectins. II. Analysis by stepwise enzymic degradation. J. Biol. Chem. **199**, 641.

LARSON, B. L., and K. A. GILLIS, 1953: Amperometric method for determining the sorption of iodine by starch. Anal. Chem. **25**, 802.

LATHE, G. H., and C. R. J. RUTHVEN, 1956: The separation of substances and estimation of their relative molecular sizes by the use of columns of starch in water. Biochem. J. **62**, 665.

LAUGHNAN, J. R., 1953: The effect of the sh_4 factor on carbohydrate reserve in the mature endosperm of maize. Genetics **38**, 485.

LEWIS, B. A., and F. SMITH, 1957: The heterogeneity of polysaccharides as revealed by electrophoresis on glass-fibre paper. J. Amer. Chem. Soc. **79**, 3929.

LINDEMANN, E., 1951: Fat determination and the influence of extraction on the properties of starch. Die Stärke **3**, 141.

LITTLE, R. R., 1957: Permanent staining with iodine vapor. Stain Technol. **32**, 7.

LOEWUS, F. A., and D. R. BRIGGS, 1957: A potentiometric study of the change in iodine binding capacity of amylose while retrograding in dilute solution. J. Amer. Chem. Soc. **79**, 1494.

LOHMAR, R. L., 1954: The structure of α-amylase modified waxy corn starch. J. Amer. Chem. Soc. **76**, 4608.

LURIA, S. E., 1953: General virology. Wiley (New York).

MACHLIS, L., and J. G. TORREY, 1956: Plants in action. Freeman (San Francisco).

MACMASTERS, M. M., 1953: The return of birefringence to gelatinized starch granules. Cereal Chem. **30**, 63.

MACWILLIAM, I. C., and E. G. V. PERCIVAL, 1951: The constitution of barley starch. J. Chem. Soc., p. 2259.

Madison, J. H., 1956: The intracellular location of phosphorylase in tobacco (*Nicotiana tabacum* L.) Plant Physiol. **31**, 387.

Maige, A., 1935: Nouvelles observations sur l'évolution des plastes amylogènes dans les cellules à réserves d'amidon. C. R. **200**, 1618.

Marré, E., and L. Felici, 1952: Factors regulating the phosphorylase activity of leaves. Chem. Abstr. **46**, 11345 b.

Maruo, B., and T. Kobayashi, 1951: Enzymic scission of the branch links in amylopectin. Nature **167**, 606.

Maywald, E., R. Christensen, and T. J. Schoch, 1955: Development of starch and phytoglycogen in Golden Sweet Corn. Agric. a. Food Chem. **3**, 521.

Mazurs, E. G., T. J. Schoch, and F. E. Kite, 1957: Graphical analysis of the Brabender viscosity curves of various starches. Cereal Chem. **34**, 141.

McGivern, M. J., 1957: Mitochondria and plastids in sieve-tube cells. Amer. J. Bot. **44**, 37.

Melamed, M. D., 1957: Unpublished results (Dept. of Botany, University of the Witwatersrand, Johannesburg).

Mes, M. G., and I. Menge, 1954: Potato shoot and tuber cultures *in vitro*. Physiol. Plantarum **7**, 637.

Meyer, A., 1895: Untersuchungen über die Stärkekörner. Fischer (Jena).

*Meyer, K. H., 1952: Starch degradation by amylases. Bakers Digest **26**, 104.

*— 1952: The past and present of starch chemistry. Experientia **8**, 405.

— et R. Menzi, 1953: Sur la structure fine de l'amidon. Helv. **36**, 702.

— et W. Settele, 1953: Sur l'inhomogéneité des amylopectines de tapioca et de waxy maize. Helv. **36**, 197.

Mikus, F. F., R. M. Hixon, and R. E. Rundle, 1946: The complexes of fatty acids with amylose. J. Amer. Chem. Soc. **68**, 1115.

Miller, E. V., 1953: Within the living plant. Blakiston (New York).

Mitchell, W. A., and E. Zillmann, 1951: The effect of fatty acids on starch and flour viscosity. Transact. Amer. Assoc. Cereal Chem. **9**, 64.

Monod, J., et A. M. Torriani, 1948: Synthèse d'un polysaccharide du type amidon aux dépens de maltose, en présence d'un extrait enzymatique d'origine bactérienne. C. R. **227**, 240.

Mould, D. L., and R. L. M. Synge, 1952: Electrokinetic ultrafiltration analysis of polysaccharides. Analyst **77**, 964.

Murthy, H. B. N., and M. Swaminathan, 1956: *In vitro* digestibility studies on banana stem starch. Biol. Abstr. **30**, 979. Nr. 9979.

— G. R. Rao, and M. Swaminathan, 1957: Studies on the starch-synthesizing enzymes in tapioca (*Manihot utilissima*) roots. Enzymol. **28**, 63.

Nakamura, M., 1953: Effect of added primer on lima bean phosphorylase. Nature **171**, 795.

— 1954: The enzymic formation and degradation of starch. Chem. Abstr. **48**, 757 e.

Nelson, C. D., and P. R. Gorham, 1957: Translocation of radioactive sugars in the stems of soybean seedlings. Canad. J. Bot. **35**, 703.

Neufield, E. F., and W. Z. Hassid, 1955: Hydrolysis of amylose by β-amylase and Z-enzyme. Arch. Biochem. Biophys. **59**, 405.

Nikuni, Z., and S. Hizukuri, 1957: Structure of starch granules. Sympos. on starch and amylases, Osaka, p. 35; Mem. Inst. Sci. Ind. Res. Osaka Univ. **14**, 173.

— and R. L. Whistler, 1957: Unusual structures in corn starch granules. J. Biochem. **44**, 227.

— H. Fuwa, and K. Takaoka, 1953: The specific change of iodine reaction on sweet-potato starch caused by some micro-organisms. Chem. Abstr. **47**, 6459 i.

Norberg, E., and D. French, 1950: Redistribution reactions of macerans amylase. J. Amer. Chem. Soc. **72**, 1202.

Norstog, K. J., 1956: Growth of rye-grass endosperm *in vitro*. Bot. Gaz. **117**, 253.

Nussenbaum, S., 1951: Differentiation of amylopectin, amylodextrins and amylose-fatty acid complexes. A spot test. Anal. Chem. **23**, 1478.

— and W. Z. Hassid, 1952: Mechanism of amylopectin formation by the action of Q-enzyme. J. Biol. Chem. **196**, 785.

Obata, Y. *et al.* 1955; 1957: The inhibitor for starch accumulation in the root of *Beta vulgaris*. Chem. Abstr. **49**, 8397 e; **51**, 16605 g.

O'Brien, J. A., 1951: Plastid development in the scutellum of *Triticum aestivum* and *Secale cereale*. Amer. J. Bot. **38**, 684.

O'Colla, P. S., and E. Lee, 1956: Synthetic polysaccharides. Chem. a. Ind. p. 522.

OHASHI, K., 1957: Properties of various starches. Chem. Abstr. 51, 11746 e.

— T. HAYAKAWA, and T. TAKAHASHI, 1957: Hygroscopicity of various kinds of starch and its application as a desiccating agent. Chem. Abstr. 51, 13431 h.

ONO, H., 1956: Intracellular conditions which control the interchange of starch and sugar in plants. Sieboldia 1, 171.

— and Y. KONAGAMITSU, 1956: The formation of starch in the isolated chloroplast. Bot. Mag. (Tokyo) 69, 146.

— — 1957: On the apparent relationship between acid phosphatase activity and starch content in plant tissues. Sieboldia 2, 1.

ONO, S., S. TSUCHIHASHI, and T. KUGE, 1953: On the starch-iodine complex. J. Amer. Chem. Soc. 75, 3601.

ONODERA, K., et al., 1953: The mechanism of formation of polysaccharides. Chem. Abstr. 47, 8786 b.

PAECH, K., und E. KRECH, 1953: Über die Stärkebildung in den Plastiden. Planta 41, 391.

PAZUR, J. H., and T. BUDOVICH, 1955: Hydrolysis of amylotriose by crystalline salivary amylase. Science 121, 702.

— and R. M. SANDSTEDT, 1954: Identification of the reducing sugars in amylolyzates of starch and starch oligosaccharides. Cereal Chem. 31. 416.

PEAT, S., J. R. TURVEY, and J. M. EVANS, 1957: Isolation of nigerose from Floridean starch. Nature 179, 261.

— W. J. WHELAN, and J. M. BAILEY, 1953: The minimum chain length for Q-enzyme action. J. Chem. Soc., p. 1422.

— — and G. JONES, 1957: Structural requirements of D-enzyme with respect to acceptors. J. Chem. Soc., p. 2490.

— — and G. W. F. KROLL, 1956: The dextrins synthesized by D-enzyme. J. Chem. Soc., p. 53.

— — and S. J. PIRT, 1949: The amylolytic enzymes of soya bean. Nature 164, 499.

— — and W. R. REES, 1956: The disproportioning enzyme (D-enzyme) of the potato. J. Chem. Soc., p. 44.

— — and G. J. THOMAS, 1952: Evidence of multiple branching in waxy maize starch. J. Chem. Soc., p. 4546.

— — — 1956: Evidence of multiple branching in waxy-maize starch. A correction. J. Chem. Soc., p. 3025.

— — P. N. HOBSON, and G. J. THOMAS, 1954: The action of R-enzyme on glycogen. J. Chem. Soc., p. 4440.

PETROVA, A. N., 1952 a: The muscle enzyme which splits the 1.6-bond in polysaccharides. Chem. Abstr. 46, 2103 i.

— 1952 b: Some properties of amylose isomerase. Chem. Abstr. 46, 7603 f.

PFAHLER, P. L., H. H. KRAMER, and R. L. WHISTLER, 1957: Effects of genes on birefringence end-point temperature of starch grains in maize. Science 125, 441.

PFANMULLER, J., and A. NOE, 1952: Enzymatic synthesis of higher carbohydrates from dextrose. Science 115, 240.

PLAISTED, P. H., 1957: Growth of the potato tuber. Plant Physiol. 32, 445.

PORTER, H. K., 1950: Inhibition of plant phosphorylases by β-amylase and detection of phosphorylase in barley. Biochem. J. 47, 476.

— 1953: The inhibition of α-amylase and phosphatase contaminants of potato phosphorylase. J. Exper. Bot. 4, 44.

— and L. H. MAY, 1955: Metabolism of radioactive sugars by tobacco leaf discs. J. Exper. Bot. 6, 43.

— and W. R. REES, 1954: Some effects of ethanol extracts of potatoes on the activity of a phosphorylase preparation. Plant Physiol. 29, 514.

POTTER, A. L., and W. Z. HASSID, 1951: The molecular constitution of amylose subfractions. J. Amer. Chem. Soc. 73, 593.

— V. SILVEIRA, R. M. MCCREADY, and H. S. OWENS, 1953: Fractionation of starches from smooth and wrinkled seeded peas. Molecular weights, end-group assays and iodine affinities of the fractions. J. Amer. Chem. Soc. 75, 1335.

PREECE, I. A., 1957: Cereal carbohydrates. The Roy. Inst. of Chem. Lectures, Monographs and Reports, No. 2, p. 8—9.

RADWAN, M. A., 1957: The use of C^{14} uniformly labelled glucose in the study of starch synthesis in the leaf. Arch. Biochem. Biophys. 68, 467.

— and C. R. STOCKING, 1957: The isolation and characterization of sunflower leaf starch. Amer. J. Bot. 44, 682.

Ram, J. S., and K. V. Giri, 1952: Starch-synthesizing enzymes of Green Gram (*Phaseolus radiatus*). Arch. Biochem. Biophys. 38, 231.

Rao, P. S., and R. M. Beri, 1955: *Dioscorea* starches. Chem. Abstr. 49, 12024 d.

Rhoades, M. M., and A. Carvalho, 1944: The function and structure of the parenchyma sheath plastids of the maize leaf. Bull. Torrey Bot. Club 71, 335.

Roberts, E. A., and B. E. Proctor, 1954: The appearance of starch grains of potato tubers of plants grown under constant light and temperature conditions. Science 119, 509.

Röbbelen, G., 1957: Eine Blattfarbmutante ohne Chlorophyll b von *Arabidopsis thaliana* (L.) Heynh. Naturw. 44, 288.

Rozenfeld, E. L., and E. G. Physhevskaya, 1954: The effect of structure of different polysaccharides of vegetable and animal origin on their capacity to form complexes with proteins. Chem. Abstr. 48, 9423 h.

Rundle, R. E., and D. French, 1943: Optical properties of crystalline starch fractions. J. Amer. Chem. Soc. 65, 558.

Rutgers, R., 1955: Starch and vanillin. J. Sci. Food Agric. 6, 735.

Samec, M., und J. R. Katz, 1933: Zur Einteilung der Stärkearten in Gruppen nach dem Röntgenspektrum und nach den Eigenschaften des Amylopektins. Z. physik. Chem. 163 (A), 291.

Sandstedt, R. M., 1955: Photomicrographic studies of wheat starch. III. Enzymatic digestion and granule structure. Cereal Chem. (Suppl.) 32, No. 3.

— 1957: Films on studies of starch gelatinization, digestion and structure. Sympos. on Starch and Amylases, Osaka, p. 54.

— and R. L. Gates, 1954: Raw starch digestion: a comparison of the raw starch digesting capabilities of the amylase systems from four α-amylase sources. Food Res. 19, 190.

Sanyal, P. K., and T. E. Wallis, 1957: The structure of rhizomes and roots of *Valeriana officinalis* and its polyploid forms. Chem. Abstr. 51, 8896 c.

*Schoch, T. J., 1945: The fractionation of starch. Adv. Carbohyd. Chem. 1, 247.

— and A. L. Elder, 1953: Starches in the food industry. Adv. in Chem. Series, p. 21.

— and D. French, 1947: Studies on bread staling. I. The role of starch. Cereal Chem. 24, 231.

— and E. C. Maywald, 1956: Microscopic examination of modified starches. Anal. Chem. 28, 382.

Schwarz, W., 1931: Entwicklungsgeschichte der Plastiden einiger grüner Pflanzen. Z. Bot. 25, 1.

Schwimmer, S., 1957: Phosphorylase inhibitor in potato: separation from activator and possible relation to chlorogenic acid. Nature 180, 149.

— and A. K. Balls, 1949: Starches and their derivatives as adsorbents for malt α-amylase. J. Biol. Chem. 180, 883.

— and W. J. Weston, 1956: Effect of phosphate and other factors in potato extract on amylose formation by phosphorylase. J. Biol. Chem. 220, 143.

Seybold, A., 1940: Zur Physiologie des Chlorophylls. S. B. Heidelberger Akad. Wissensch., p. 1.

Shaw, M., 1954: Phosphorylase in the chloroplasts of wheat. Canad. J. Bot. 32, 523.

Singh, S., N. Nath, and H. P. Nath, 1956: Iodine-binding capacity of some native starches of Indian food grains. Biochem. J. 63, 718.

Sisakyan, N. M., 1954: Die fermentative Aktivität der protoplasmatischen Strukturen. Akademie-Verlag (Berlin).

*Stacey, M., 1954: Enzymic synthesis of polysaccharides. Adv. Enzymol. 15, 301.

Stacy, C. J., and J. F. Foster, 1956: Flow birefringence behaviour of some amylopectins and limit detrins. J. Polymer Sci. 20, 67.

— — 1957: Molecular-weight heterogeneity in starch amylopectins. J. Polymer Sci. 25, 39.

Stafford, H. A., 1951: Intracellular localization of enzymes in pea seedlings. Physiol. Plantarum 4, 696.

Stepanenko, B. N., and E. M. Afanas'eva, 1957: Structure and the iodine reaction of amylopectins and of crystalline amylose of potato tubers during their maturation under different conditions of cultivation. Chem. Abstr. 51, 11485 f.

Steward, F. C., and S. M. Caplin, 1951: A tissue culture from potato tuber: the synergistic action of 2, 4-D and of coconutmilk. Science 113, 518.

Steward, F. C., R. G. S. Bildwell, and E. W. Yemm, 1956: Protein metabolism, respiration and growth: a synthesis of results from the use of ^{14}C-labelled substrates and tissue cultures. Nature 178, 734, 789.

Stocking, C. R., 1952: The intracellular location of phosphorylase in leaves. Amer. J. Bot. 39, 283.

— 1956: Precipitation of enzymes during isolation of chloroplasts in carbowax. Science 123, 1032.

Strandine, E. J., G. F. Carlin. G. A. Werner, and R. P. Hopper, 1951: Effect of monoglycerides on starch, flour, and bread: a microscopic and chemical study. Cereal Chem. 28, 449.

Straus, J., 1954: Maize endosperm tissue grown in vitro. II. Morphology and cytology. Amer. J. Bot. 41, 833.

— and C. D. La Rue, 1954: Maize endosperm tissue grown in vitro. I. Culture requirements. Amer. J. Bot. 41, 687.

Stuart, W. N., 1948: A study of the plastids in the cells of the mature sporophyte of Isoetes. Bot. Gaz. 110, 281.

Stuart, N. W., and B. Brimhall, 1943: Starch from Easter Lily bulbs. Cereal Chem. 20, 734.

Subrahmanyan, V., G. Lal, D. S. Bhatia, N. L. Jain, G. S. Bains, K. V. Srinath, B. Anandaswamy, B. H. Krishna, and S. K. Lakshminarayana, 1957: Studies on banana pseudostem starch: Production, yield, physico-chemical properties and uses. J. Sci. Food Agric. 8, 253.

Summer, R., and D. French, 1956: Action of β-amylase on branched oligosaccharides. J. Biol. Chem. 222, 469.

Sveshnikova, D. I. N., 1956: Formation of fats in starch grains and plastids. Research 9, 396.

Tagawa, T., et al., 1956: Physiological and morphological studies on potato plants. Chem. Abstr. 50, 12207 a.

Takaoka, K., and J. Nikuni, 1953: The influence of fatty acids on the iodine-colour reaction of starch paste. Chem. Abstr. 47, 4939 f.

Talwar, G. P., E. Barber, J. Basset, et M. Mácheboeuf, 1951: Etudes sur l'hydrolyse enzymatique de l'amidon sous des pressions élevées. Bull. Soc. Chim. Biol. 33, 1793.

Tanaka, K., 1954: Enzymatic studies on the mechanism of polysaccharide formation in maize seed. I. Primer specificity of P-enzyme. Mem. Osaka Univ. 3 (B), 99.

— 1955, 1956: Q-enzyme of non-waxy maize seeds. Bot. Mag. (Tokyo) 68, 352; 69, 281.

Täufel, K., G. Feldmann, und W. Panzer, 1956: Pektingelierung und Dehydratation. Naturw. 43, 374.

Thornburg, W. L., 1956: An electron microscopic examination of the formation of starch granules in corn endosperm. Ph. D. thesis, Purdue University.

Ueda, S., A. Shimizu, and I. Ota, 1955: The relations between the maturity of paddy rice grain and the physicochemical characters of their starches. Chem. Abstr. 49, 7877 f.

Ulmann, M., 1953: Die Grundlagen des thermischen Stärkeabbaues. Kolloid-Z. 130, 31.

— 1954: Die verschiedenen Strukturelemente der Stärke. Die Stärke 6, 95.

— 1956 a: Chromatographic investigations of acid hydrolysis of potato starch. Chem. Abstr. 50, 16146 g.

— 1956 b: Über die äußere Hülle des Stärkekorns. Die Stärke 8, 109.

— 1957: Bestimmung der chemischen Natur der Hülle eines gequollenen Stärkekorns. Kolloid-Z. 150, 128.

Umrau, A. M., and F. Smith, 1957: A chemical method for the determination of the molecular weight of certain polysaccharides. Chem. a. Ind., p. 330.

Vittorio, P. V., G. Krotkov, and G. B. Reed, 1954: Synthesis of radioactive sucrose by tobacco leaves from C^{14} uniformly labelled glucose and glucose-1-phosphate. Canad. J. Bot. 32, 369.

Volz, F. E., and P. E. Ramstad, 1952: Effects of various physical treatments upon the amyloclastic susceptibility of starch. Food Res. 17, 81.

Wanner, H., 1952: Phosphataseverteilung und Kohlehydrattransport in der Pflanze. Planta 41, 190.

Weber, F., 1920: Notiz zur Kohlensäureassimilation von Neottia. Ber. dtsch. bot. Ges. 38, 233.

Wettstein, D. von, 1955: Formation of the plastid structure as affected by mutation in the chlorophyll apparatus. Fine Structure of Cells, a Symposium. Noordhoff (Groningen), p. 55.
*Whelan, W. J., 1953: The enzymic breakdown of starch. Biochem. Soc. Symposia, Cambridge, 11, 17.
— and J. M. Bailey, 1954: The action pattern of potato phosphorylase. Biochem. J. 58, 560.
— and P. J. P. Roberts, 1952: Action of salivary α-amylase on amylopectin and glycogen. Nature 170, 748.
Whistler, R. L., and W. L. Thornburg, 1957: Development of starch granules in corn endosperm. Agric. Food Chem. 5, 203.
— and E. S. Turner, 1955: Fine structure of starch granule sections. J. Polymer Sci. 18, 153.
— and P. Weatherwax, 1948: Amylose content of Indian corn starches from North, Central, and South American corns. Cereal Chem. 25, 71.
— J. D. Byrd, and W. L. Thornburg, 1955: Surface structure of starch granules. Biochem. Biophys. Acta 18, 146.
— H. H. Kramer, and R. D. Smith, 1957: Effect of certain genetic factors on the sugars produced in corn kernels at different stages of development. Arch. Biochem. Biophys. 66, 374.
Whittenberger, R. T., and G. C. Nutting, 1948: Potato-starch gels. Ind. Eng. Chem. 40, 1407.
Witnauer, L. P., F. R. Senti, and M. D. Stern, 1952: Molecular weight of potato amylopectin as determined by light scattering. J. Chem. Phys. 20, 1978.
Wolf, M. J., M. M. MacMasters, J. E. Hubbard, and C. E. Rist, 1948: Comparison of corn starches at various stages of kernel maturity. Cereal Chem. 25, 312.
Wolff, I. A., H. A. Davis, J. E. Cluskey, L. J. Gundrum, and C. E. Rist, 1951: Preparations of films from amylose. Ind. Eng. Chem. 43, 915.
— B. T. Hofreiter, P. R. Watson, W. L. Deatherage, and M. M. MacMasters, 1955: The structure of a new starch of high amylose content. J. Amer. Chem. Soc. 77, 1654.
Wolfrom, M. L., and A. Thompson, 1956: Occurrence of the (1 ⟶ 3)-linkage in starches. J. Amer. Chem. Soc. 78, 4116.
— — 1957: Degradation of glycogen to isomaltotriose and nigerose. J. Amer. Chem. Soc. 79, 4212.
Wunderly, C., 1940: Die Steuerung eines enzymatischen Abbaues durch einen anderen. Helv. 23, 414.
Wyk, A. J. A. van der, et J. Schmorak, 1953: Sur l'existence de celluloses ramifiées. Viscométrie comparative de solutions de celluloses et d'amidons. Helv. 36, 25.
Yin, H. C., and C. N. Sun, 1949: Localization of phosphorylase and of starch formation in seeds. Plant Physiol. 24, 103.
Ziegler, H., 1956: Untersuchungen über die Leitung und Sekretion der Assimilate. Planta 47, 447.
Zimmermann, M. H., 1957: On the translocation mechanism in the phloem of White Ash (Fraxinus americana L.). Plant Physiol. 32, 399.